T0139282

Python Programming

Python Programming

A Practical Approach

Vijay Kumar Sharma

Vimal Kumar

Swati Sharma

Shashwat Pathak

CRC Press
Taylor & Francis Group
Boca Raton London New York

CRC Press is an imprint of the
Taylor & Francis Group, an **informa** business

First edition published 2022
by CRC Press
6000 Broken Sound Parkway NW, Suite 300, Boca Raton, FL 33487-2742

and by CRC Press
2 Park Square, Milton Park, Abingdon, Oxon OX14 4RN

© 2022 Taylor & Francis Group, LLC

CRC Press is an imprint of Taylor & Francis Group, LLC

Library of Congress Cataloging-in-Publication Data
A catalog record has been requested for this book

ISBN: 978-1-032-02849-1 (hbk)
ISBN: 978-1-032-02852-1 (pbk)
ISBN: 978-1-003-18550-5 (ebk)

DOI: 10.1201/9781003185505

Typeset in Palatino
by Newgen Publishing UK

We dedicate this book to my grandmother

Contents

Figures

Tables

Preface

The impact of a technology largely depends upon its quick and easy deployment, ease of use and seamless integration. Over the years, Python has translated into a high-level programming language, which has influenced the deployment of emerging technologies such as machine learning, artificial intelligence, block chain, and deep learning, etc., significantly.

Python is a general-purpose, high-level programming language. It is open source software and its source code is available with a license in which the copyright holder provides the right to study, change, and distribute the software to anyone and for any purpose. Python is perfect for data analysis, artificial intelligence, and scientific computing. Instagram and YouTube use software written in Python. It is used in numerous application domains. The Python standard library supports many internet protocols like HTML, XML, JSON, FTP, IMAP etc.

In businesses, Python is used in domains such as medical, aviation, industrial, financial services, GPS, and consumer goods industry etc. Governments utilize it for administration, homeland security, traffic control, public safety etc. In network programming, Python is used to control firmware updates. It plays a vital role in computer graphics, data mining, embedded systems, etc.

This book smoothly takes you from the basics of Python to advanced areas. This book is a golden key for Python enthusiasts, students, and researchers.

Salient Features

- Based on real-world requirements and solutions.
- Simple presentation without avoiding necessary details of the topic.
- Executable programs on almost every topic.
- Plenty of exercise questions, designed to test students' skills and understanding.

How this book is different?

With ample number of books on Python already available in the market, why read this book?

The simple truth is none of the books has really worked for the beginner who has no prior knowledge about programming. We personally believe that a textbook must identify potential areas of confusion and anticipate the questions that a reader is likely to come up.

We put ourselves at the student's desk, pose their questions, and answer them as well as we can. This book is primarily meant for self-study even though some guidance at times may be helpful. Extensive cross-referencing has been made available by embedding parenthesized section numbers in the text.

Solved examples within the chapters demonstrate the applicability of the concepts discussed. Programming assignments at the end of each chapter facilitate revision, with notes highlighting the important terms and concepts.

Content and Structure

A brief description of each chapter is given below:

Chapter 1 deals with **Introduction to Python Programming** covering features, limitations, flavors, application areas, how to run, internal working and comments in Python.

Chapter 2 covers the **Basics of Python Programming** including identifiers, reserved keywords, literals, fundamental data types, base conversion, type casting, escape characters, math module, input, evaluate, print function, command line arguments, and the delete statement.

Chapter 3 covers the different types of **Operators in Python** and operator precedence in Python.

Chapter 4 includes the **Control Flow in Python**, which comprises conditional statements, iterative statements, and transfer statements.

Chapter 5 comprises the detailed concept of **Strings**.

Chapter 6 explains different **Data Structures of Python** such as list, tuple, set, and dictionaries.

Chapter 7 deals with **Functions** in Python including types of function, arguments in function, scopes of variables, global keyword, recursive function, anonymous function, lambda function, filter function, map function, reduce function, and the nested function.

Chapter 8 describes **Modules** processing including member aliasing, module aliasing, reloading a module, working with math module, working with random module and packages.

Chapter 9, 10 deal with complete **Object-Oriented Programming**.

Chapter 11 includes **Exception Handling**.

Chapter 12 deals with **File Handling**, which includes types of files, opening, closing a file, various properties of file, writing data to text files, handling binary data, handling CSV files and all the detailed concept about file handling.

Chapter 13 describes **Multithreading** and multitasking concepts.

Chapter 14 deals with **Synchronization**, semaphore, inter thread communication, types of queues, and other related concepts.

Chapter 15 deals with **Regular Expressions and Web Scraping**.

Chapter 16 describes **Python Database Programming** in detail.

Acknowledgements

First and foremost, the authors would like to thank their teachers, including Prof. Rajeev Tripathi, Director, MNNIT Allahabad, Prayagraj, for his invaluable research inputs, incisive comments, valuable moral support, and kind help during the course of this book project. Needless to say, without his help and support, this book would not have been possible. We would like to express our deepest gratitude to him. Dr. R.S. Yadav (Professor), Dr. D.K. Yadav (Professor) and Dr. Dinesh Singh (Asst. Professor), Dept. of CSE and Dr. Basant Kumar (Associate Professor) Dept. ECE, MNNIT Allahabad, Prayagraj for all their continuous support, invaluable advice, guidance, freedom, patience, and protection to do research without pressure. We have learnt a lot from them on general methodology and specific structure related to this book. They always encouraged us, prevented us from getting stymied by the obstacles en route and helped us in every possible way. They have perfect writing skills, and creative thinking that impressed us deeply. Particularly, they gave us invaluable suggestions and timely support on this project. Our choice of career path has been greatly influenced by their suggestions, and we hope that we can live up to their high standards. We really owe them a lot for all their painstaking efforts. Their deep insight and proficiency have done much to shape this book and the analytical mind in all of us. It is our profound pleasure to place on record our hearty gratitude to these personalities.

We offer our sincere thanks to the editorial board members Ms. Aastha Sharma and Ms. Shikha Garg for providing excellent guidance during the period of finalizing the contents of this manuscript and shaping it in an interesting format. We gratefully acknowledge our seniors Dr. Pawan Kumar Verma (Asst. Professor) Dept. of ECE, NIT Jalandhar and Dr. Rajesh Verma (Professor) Dept. of Electrical Engineering, King Khalid University, Al Madinah, for clearing our doubts and sharing their ideas. We particularly thank our senior Dr. Pawan Verma for his guidance, useful suggestions, and constant encouragement during the period of writing. We are thankful for their co-operation and fruitful discussions. We would like to express our special thanks to our friends for understanding and solving various personal problems. Their support and understanding gave us strength during the course of this work. The happy times, their funny activities, fights, and the moments that we shared with them are memorable for us throughout lifetime. Without them, this book would not be possible.

We owe a special debt of gratitude to our parents and family. They have, more than anyone else, been the reason we have been able to get this far. Words cannot express our gratitude to our parents, siblings and spouses who extended their unconditional support, love, blessings, and inspiration and have made this effort successful. They instilled in us the value of hard work and taught to overcome life's disappointments. Sincere thanks to our better halves Dr. Reena (wife of Dr. Kumar), Ms. Nandita Upadhyay (wife of Dr. Pathak), Mr. Pankaj Kumar Sharma (husband of Dr. Swati Sharma) for their patience and constant support both morally and professionally. Mr. Vijay Kumar Sharma wants to specially thank his father (Mr. Mithai Lal Sharma (Royal Tiger)) and mother (Mrs. Malti Sharma) for constant support and motivation.

Last, but not least, we would like to express our heartiest gratitude towards our forefathers and God Almighty for taking care of all the blessings and learning experiences.

About the Authors

 **Vijay Kumar Sharma** is an Assistant Professor in the Department of Computer Science & Engineering at MIET, Meerut, (U.P), India. He received his B.Tech degree in 2012 from Uttar Pradesh Technical University, Lucknow, and M.Tech degree in 2017 from Motilal Nehru National Institute of Technology, Allahabad, India. He has published various research papers in international and national journals and conferences of high repute. His research interests lies in the area of cloud computing, artificial intelligence and blockchain technology. Various achievements in the programming field like winner in 'Coding Competition Zonal Level 2019', winner in 'Smart India Hackathon 2019', winner of 'India-EU-ICT Smart City Hackathon 2019 Pune'.

 **Dr. Vimal Kumar** is an Associate Professor in the Department of Computer Science & Engineering at MIET, Meerut, (UP), India. He received his B.Tech degree in 2007 from Uttar Pradesh Technical University, Lucknow, and M.Tech degree in Information Security from Motilal Nehru National Institute of Technology, Allahabad, India in 2011. He did his Ph.D. in Computer Science and Engineering at MMMEC, Gorakhpur (AKTU, Lucknow), India, in 2017. He has published a large number of research papers in international and national journals and conferences of high repute. His research interests lie in mobile ad hoc network, network security and network forensics.

 **Dr. Swati Sharma** did her graduation in Information Technology with Honors degree in 2010 and M.Tech with Honors degree in 2015. She did her Ph.D. in Computer Science and Engineering. An academician with more than ten years of teaching experience, she has more than a dozen of research papers in reputed Scopus indexed journals, international journals, and IEEE conferences to her credit. Her areas of interest are data mining, data analysis, algorithm analysis and design. She is currently doing research in the area of data analysis. She has national level certifications on Python and R programming language. She is certified from IBM Db2, RAD, RSA and RTC. She is the prime author of the book *Neural Network and Fuzzy Time Series* with the international publisher Lambert Academic Publishing.

Dr. Shashwat Pathak did his graduation in Electronics and Communication Engineering in the year 2009. Later, he completed his Master's in Communication Technology from the Department of Electronics and Communication, University of Allahabad, Prayagraj, India, in 2012. He earned his Ph.D. from the Department of Electronics and Communication Engineering, Motilal Nehru National Institute of Technology Allahabad, Prayagraj, India (MNNIT Allahabad), in December 2017. His research interests includes wireless communication, telemedicine systems and services, digital image processing, and designing medical diagnostics and assisting devices for patients. He is the head of the Incubation Centre, IIT Patna, funded by the Ministry of Electronics and IT (MeiTY), GoI for his thesis on "Portable and Automated Cataract Detection and Grading System," He is co-founder and MD of Electro Curietech Pvt. Ltd., which works in the area of medical electronics, telemedicine solutions, office IT solutions, and consultancy on sustainability. He is center-head of the Atal Community Innovation Center MIET Meerut Foundation, where he is has great involvement in nurturing students and nearby communities to create impact-making innovations and ventures for social causes.

1

Introduction to Python Programming

1.1 Introduction

Python is a high-level programming language. The father of Python is Guido Van Rossum; he created Python in 1989, while he was working at the National Research Institute at Netherlands, but formally it came into the picture in 1991. It is the most popular programming language. He provided Python with the features and useful characteristics of almost all programming languages such as the functional programming feature of C, object oriented features of C++, modular programming of Modula-3, scripting language of Perl and many more.

Program: Write a program to print Hello Python.

In Java:

```java
public class Test
  {
    public static void main(String[] args)
      {
         System.out.println ("Hello Python");
      }
  }
```

In C:

```c
#include<stdio.h>
void main()
  {
     print("Hello Python");
  }
```

In Pascal:

```pascal
program Hello;
begin
     Writeln('Hello Python')
End
```

DOI: 10.1201/9781003185505-1

In Python:

print('Hello Python')

1.2 Importance

1. **User friendly**: Python is very easy to learn. Its programs are clearly defined and easily readable. It uses few keywords and a clearly defined syntax. This makes it easy for just anyone to pick up the language quickly.
2. **Freeware:** Python is freeware software. Therefore, anyone can freely use the code to write new programs.
3. **High-level programming language**: Programmers just have to deal with the solution of the problem; they do not have to think about low-level details.
4. **Platform independent:** The Python virtual machine converts the code into machine understandable form once written, so it is platform independent.
5. **Portability:** Python is a portable language and hence the programs run the same on any platform.
6. **Dynamically typed:** Python is considered as a dynamically typed language, as there is no need to declare the data type of the variable used. It automatically takes its data type whereas in Java, C, etc. the data types need to be declared as they are statically typed languages.
7. **Procedural and object-oriented:** Python aids an object-oriented as well as procedure oriented style of programming. While the object-oriented technique encapsulate data and functionalities within objects, the procedure-oriented technique, on the other hand, builds the program around procedures or functions that are nothing but reusable pieces of programs.
8. **Interpreted:** There is no need to compile a program before executing it as it is compiled at run time only. You can simply run the program (Jenkins 2004).
9. **Extensible:** Since Python is open source software, anyone can add low-level modules to the Python interpreter (Zhang 2015). These modules enable programmers to add or customize their tools to work more efficiently.
11. **Vast library:** Python provides a rich in-built library, which can be used directly by the programmer.

1.3 Limitations of Python

1. Instead of a compiler, Python executes with the help of an interpreter, which results into a slight bit slowdown of the execution process.
2. It is not recommended for mobile platforms.

1.4 Python Impressions

1. **CPython:** It is the basic impression, which is used to process C language.
2. **Jython OR JPython:** Useful for the Java platform and can be executed on JVM.
3. **IronPython:** Useful for C#.Net applications.
4. **PyPy:** PyPy is used for increasing the performance as Just-In-Time compilers are supported by PVM.
5. **Ruby Python:** Useful for Ruby-based applications
6. **Anaconda Python:** Useful for steering a tremendous amount of data processing.

1.5 How to Run Python

Python can be run in the following ways:

1.5.1 From the Command Line

This method invokes the interpreter using a script framework, which starts executing the script and keep on executing until it successfully completes the execution. On completion of the execution of script, the interpreter diminishes. Python script can be processed at the command line by invoking the interpreter (Gowrishankar and Veena 2018). In Figure 1.1, the Python command prompt contains an opening message >>>, called the command prompt. The cursor at the command prompt waits for the user to enter the Python command.

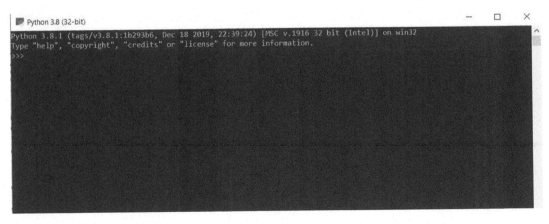

FIGURE 1.1
Python interactive mode as Python command line window

A simple command has been executed at the command prompt in Figure 1.2.

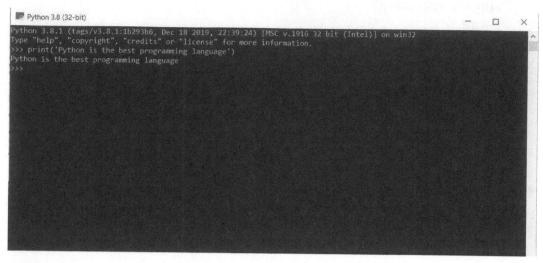

FIGURE 1.2
Simple command executed in Python command line

1.5.2 From the Integrated Development Environment (IDLE)

One can also run Python from a graphical user interface, if you have a GUI application on your system that supports Python (Van Rossum 2007). In Figure 1.3, the Python interactive shell prompt contains an opening message >>>, called a shell prompt. The cursor at the shell prompt waits for the user to enter a Python command.

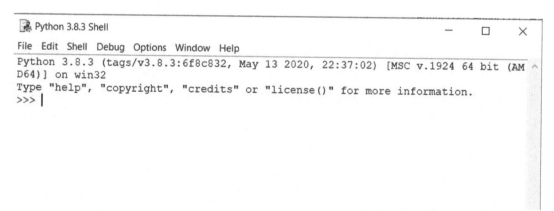

FIGURE 1.3
Python IDLE – interactive shell

Figure 1.4 shows simple commands that are executed in the interactive mode, i.e. the interactive shell of Python IDLE.

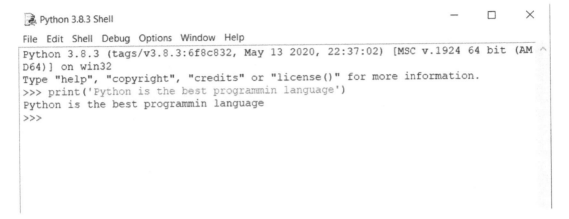

FIGURE 1.4
Simple command executed in Python IDLE

1.6 Internal Working of Python

Whenever a user tries to run Python code on the Python IDLE or Python command prompt, then various operations are being performed internally. These internal steps are broken down into three steps as shown below:

Step 1:

The interpreter reads the Python code or instruction. Then, it verifies the syntax and formatting of instructions. If it finds any errors, it immediately stops the translation and displays an error message.

Step 2:

If there is no error, then the interpreter translates it into the equivalent form in low-level language called byte code.

Step 3:

Byte code is then transmitted to the Python virtual machine and then the output is displayed.

1.7 Comments in Python

Comments are beneficial while writing a program. It tells people what the source code is about. Comments are written for better understanding of the program. To write a comment, the hash symbol (#) is used. The Python interpreter ignores the comment. Figure 1.5 shows an example of writing a comment using the hash symbol and with no output being produced.

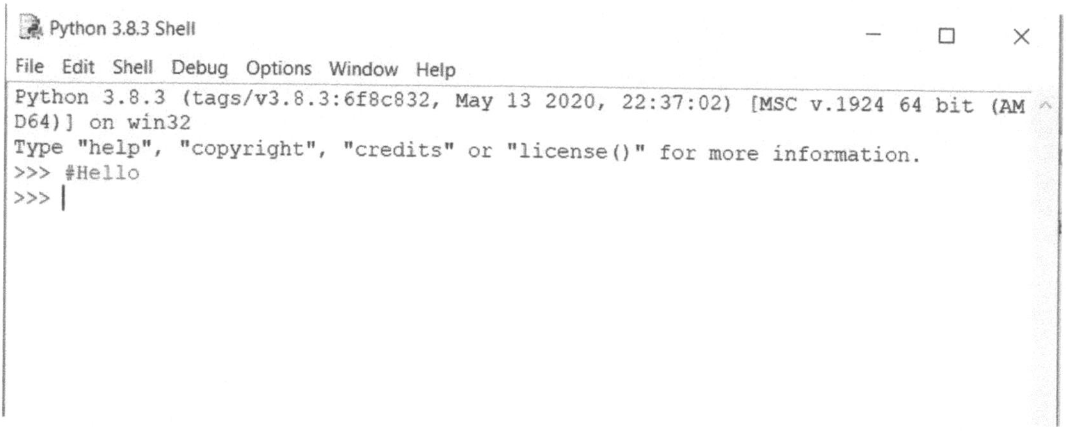

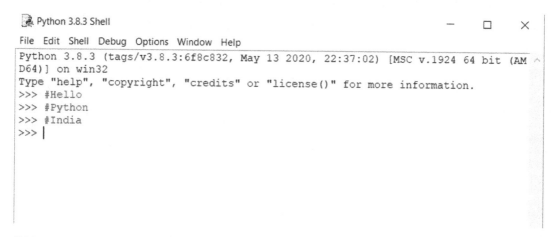

FIGURE 1.5
Showing single line comment

For multiline comments, there are two ways to write comments. The first method is to use the hash symbol at the beginning of each line. Figure 1.6 shows the use of the hash symbol for multiline comments.

FIGURE 1.6
Showing multiline comment using hash (#) symbol

The second way to write multiline comments is to use triple quotes. Figure 1.7 shows the use of triple quotes for multiline comments.

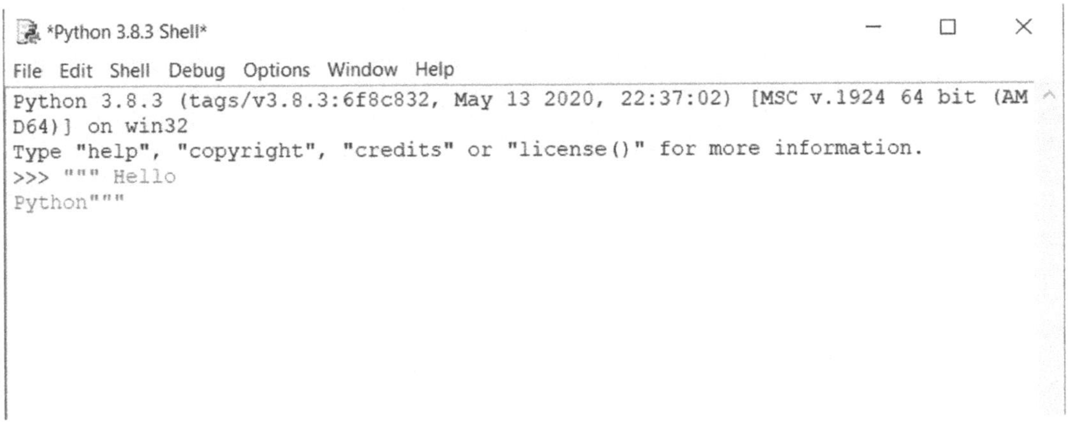

FIGURE 1.7
Showing multiline comment using triple quotes

1.8 Conclusion

In this chapter, we have presented the basic outline of Python language, its key features, application areas, features, and limitations. A brief description of running Python code on its IDE has been presented. Understanding the internal workflow of any language is a key attribute. Finally, the need to include comments to maintain good programming habits and maintain proper meaningful reference has been presented.

Review Questions

1. Is Python a case-sensitive language?
2. How can you run Python?
3. How can you write comments in Python?
4. Differentiate between command prompt and shell prompt.
5. What do you mean by the Python virtual machine?

Programming Assignments

PA 1: Write a program to display the statement, "I Love Python".
PA2: Write a program to display the following statements in two different lines.
   " I love Python"   "I love India"

References

Gowrishankar, S., and A. Veena. 2018. *Introduction to Python Programming*. CRC Press.

Jenkins, Tony. 2004. The first language-a case for python?: 1–9.

Van Rossum, Guido. 2007. Python programming language. In USENIX annual technical conference, vol. 41, p. 36.

Zhang, Yue. 2015. *An Introduction to Python and computer programming*. Springer, Singapore: 1–11.

2

Basics of Python Programming

2.1 Introduction

In order to start writing a program in Python, a user must be aware of variables, declarations, data types, libraries, functions, and their proper representation. To recognize the variables, identifiers are used, and Python has an inbuilt function to determine the class of input and output as well. Conversion of data from one base to another base and transforming the data type of a value into another data type are some common operations that are presented in this chapter. Data representation and their combination in the form of fixed and variable sets and in various other forms is explained. This chapter also discusses information regarding the escape characters, which are used in relation to the special meaning and the math module that supports various functions for performing various mathematical operations.

2.2 Identifiers

The Python identifier is used to recognize a variable, class, function, module, or any other object. Python is case sensitive and hence uppercase and lowercase letters are considered distinct.

2.2.1 Rules to Declare the Identifier

1. The only characters permitted are letters, numeric digits, and the underscore symbol(_).
   - Price = 40   ✓ (Right)
   - Pri@e = 60   ✖ (Wrong; NameError will be displayed)
2. The identifier should not start with a digit
   - 12sum   ✖ (Wrong; NameError will be displayed)
   - sum23   ✓ (Right)
3. Identifiers are case sensitive in nature.

Example:

```
a=15
A=20
print(a)
print(A)

15
20
```

2.2.2 Characteristics of the Identifier

1. If an identifier starts with underscore (_) then it is a private identifier.
2. If an identifier starts with _ID_ then it is a strongly private identifier.
3. Identifiers should not begin with digits.
4. They are case-sensitive.
5. The reserved words cannot be used as identifiers.

   Example: if = 100 (Wrong; NameError will be displayed)

6. The dollar ($) symbol is not permitted in Python.

2.3 Reserved Keywords

These are keywords reserved by the programming language and prevent the user or the programmer from using them as an identifier in a program. There are 33 reserved keywords as shown in Table 2.1.

TABLE 2.1
Python reserved keywords

Reserved keywords				
False	class	finally	is	return
None	continue	for	lambda	try
True	def	from	nonlocal	while
And	del	global	not	with
As	elif	if	or	yield
Assert	else	import	pass	
Break	except	in	raise	

Program: Write a program to show the reserved keywords in Python.

```
Python 3.8.3 Shell                                        —    □    ×

File  Edit  Shell  Debug  Options  Window  Help
Python 3.8.3 (tags/v3.8.3:6f8c832, May 13 2020, 22:37:02) [MSC v.1924 64 bit (AM
D64)] on win32
Type "help", "copyright", "credits" or "license()" for more information.
>>> import keyword
>>> keyword.kwlist
['False', 'None', 'True', 'and', 'as', 'assert', 'async', 'await', 'break', 'cla
ss', 'continue', 'def', 'del', 'elif', 'else', 'except', 'finally', 'for', 'from
', 'global', 'if', 'import', 'in', 'is', 'lambda', 'nonlocal', 'not', 'or', 'pas
s', 'raise', 'return', 'try', 'while', 'with', 'yield']
>>> |
```

2.4 Literals

Literals are numbers, strings, or characters that appear directly in a program. The programming language contains data in terms of input and output, and any kind of data can be presented in terms of value. The value can be of any type such as literals containing numbers, characters, and strings. To know the type of value in Python, there is an inbuilt method called type.

Syntax: type(value)

Program: Show the data types of the following literals.

1. 20
2. 25.5
3. 'A'
4. "PYTHON"

```
Python 3.8.3 Shell                                        —    □    ×

File  Edit  Shell  Debug  Options  Window  Help
Python 3.8.3 (tags/v3.8.3:6f8c832, May 13 2020, 22:37:02) [MSC v.1924 64 bit (AM
D64)] on win32
Type "help", "copyright", "credits" or "license()" for more information.
>>> type(20)
<class 'int'>
>>> type(25.5)
<class 'float'>
>>> type('A')
<class 'str'>
>>> type("PYTHON")
<class 'str'>
>>> |
```

Here, int refers to integer literal
float refers to floating point literal
str refers to string literal

2.5 Fundamental Data Types

All features in Python are associated with an object. It is one of the primitive elements of Python. Further, all kinds of objects are classified into types. One of the easiest types to work are the numbers, and the native data types supported by Python are string, integer, floating point numbers, and complex numbers. Table 2.2 shows the Python inbuilt data types.

TABLE 2.2
Python inbuilt data types

Data types	
int	list
float	tuple
complex	set
bool	frozenset
str	dictionary
bytes	none
bytearray	range

2.5.1 Integer Numbers

An integer is a combination of positive and negative numbers including (zero) 0. In a program, integer literals are written without commas and a leading minus sign to indicate a negative value.

Example:

```
a = 10
print(a)
print(type(a))

10
<class 'int'>
```

Integer values are shown in the following ways:

A) Decimal Form (Base 10)
  A default number system; its permitted values are 0, 1, 2, 3, 4, 5, 6, 7, 8, 9.
B) Binary Form (Base-2)
  Its permitted values are 0 and 1; the literal value are prefixed with 0b or 0B
C) Octal Form (Base-8)
  Its permitted values are 0, 1, 2, 3, 4, 5, 6, 7; the literal values are prefixed with 0o or 0O.

D) Hexadecimal Form (Base-16)
The permitted digits are 0, 1, 2, 3, 4, 5, 6, 7, 8, 9, a–f; the literal values should be prefixed with 0x or 0X

Example:

```
p=100
q=0o100
r=0X10c
s=0B1010
print(p)
print(q)
print(r)
print(s)
```

```
100
64
268
10
```

2.5.2 Floating Point Numbers

Floating point numbers consist of a whole number, decimal point, and fraction part. The length of real numbers has infinite precession, i.e. the digits in the fractional part can continue forever. Thus, Python uses floating point numbers to represent real numbers.

Example:

```
f = 13.22
print(type(f))
f1 = 13.22e4
print(f1)
```

```
<class 'float'>
132200.0
```

2.5.3 Complex Numbers

The complex numbers are expressed in the form of (a + bj), where a and b are real numbers and j is an imaginary part (Herscovics and Linchevski 1994).

Example:

```
c=0B10+8j
print(c)
```

```
(2+8j)
```

```
c=3+0B10j
print(c)
```

```
SyntaxError: invalid syntax
```

Example: (on float data type)

```
c1=30+2.6j
c2=50+3.6j
c3=c1+c2
print(c3)
print(type(c2))

(80+6.2j)
<class 'complex'>
```

Example: (to extract real and imaginary part)

```
c1=30+2.6j
print("Real Part:",c1.real)
print("Imaginary:",c1.imag)

Real Part: 30.0
Imaginary: 2.6
```

2.5.4 Boolean Type

In Python, the Boolean data type is represented as bool. It is a primitive data type with values, i.e. True or False. The true value is represented as 1 and false as 0. Moreover, when two True are added, it will produce 2 as output (True + True = 2) whereas when False is subtracted from True, it will produce 1 as output (True – False = 1).

Example:

```
b = True
print(type(b))

<class 'bool'>
```

Example:

```
x = 10
y = 20
z = x<y
print(z)

True
```

2.5.5 String Type

A string is a combination of characters enclosed within single quote or double quotes.

Example:

```
x='a'
print(type(x))

<class 'str'>
```

The multi-line string literals cannot be denoted using single quotes or the double quotes. To denote this, you should use triple single quotes (''') or triple double quotes(""")

p = ''''vimalmnnit''''
p = """"vimalmnnit""""

The triple quotes can also be used for using single quote or double quotes in the string.

'''I like "Python''' 'I like "Python'

One string can be embedded in other string

'''Python is the "best" programming language'''

Slice operator ([]) is used to extract the parts of the string. The string starts with zero index; it can be positive and negative also.

Example:

```
s="vijay"
print(s[0])
print(s[2])
print(s[-1])
print(s[10])

v
j
y

IndexError: string index out of range
```

Example:

```
s="vimal"
print(s[1:10])
print(s[1:])
print(s[:3])
print(s[:])
print(s*4)
print(len(s))

imal
imal
vim
vimal
vimalvimalvimalvimal
5
```

Example:

```
x='a'
print(type(x))

<class 'str'>
```

2.6 Base Conversion

Python comprises various inbuilt functions for base conversions; they are discussed below:

1. bin(): It is used to transform a value into binary form.

   Example:

   ```
   print(bin(30))
   print(bin(0o11))
   print(bin(0X12F))

   0b11110
   0b1001
   0b100101111
   ```

2. oct(): It is used to transform a value into octal form.

   Example:

   ```
   print(oct(30))
   print(oct(0B1100))
   print(oct(0X126))

   0o36
   0o14
   0o446
   ```

3. hex(): It is used to transform a value into hexadecimal form.

   Example:

   ```
   print(hex(380))
   print(hex(0B101011))
   print(hex(0o1632))

   0x17c
   0x2b
   0x39a
   ```

2.7 Type Casting

Type conversion is used to transform the data type of a value into another data type. There are numerous inbuilt functions used for type casting, which are discussed below:

2.7.1 Integer: Int()

This function transforms a string or a number into a whole number to an integer; it removes everything after the decimal point.

Example:

```
print(int(12.44))
print(int(True))
print(int(False))
print(int("15"))
#print(int("15.6"))    #invalid literal for int() with base 10: '15.6'
#print(int("one"))     #invalid literal for int() with base 10: 'one'
#print(int("0B1010"))  #invalid literal for int() with base 10: '0B1010'
#print(int(15+6j))      #can't convert complex to int
```

```
12
1
0
15
```

2.7.2 Floating Point: Float()

This function transforms a string into the floating point number or it is used to transform any data type into float type.

Example:

```
print(float(15))
print(float(True))
print(float(False))
print(float("15"))
print(float("15.6"))
#print(float("one"))      #could not convert string to float: 'one'
#print(float("0B1010"))   #could not convert string to float: '0B1010'
#print(float(15+6j))      #can't convert complex to float
```

```
15.0
1.0
0.0
15.0
15.6
```

2.7.3 Complex Numbers: Complex()

This function is used to transform any data type into a complex data type.
Case 1: complex(p) is used to transform p into the complex number with p as the real part and 0 as the imaginary part (Kalicharan 2015).

Example:

```
print(complex(15))
print(complex(True))
print(complex(False))
print(complex("15"))
print(complex("15.6"))
print(complex(15+6j))
#print(complex("one")) #complex() arg is a malformed string
```

```
(15+0j)
(1+0j)
0j
(15+0j)
(15.6+0j)
(15+6j)
```

Case 2: complex (p, q) is used to transform p and q into a complex number with p as the real part and q as the imaginary part.

Example:

```
print(complex(50, -2))
print(complex(True, False))

(50-2j)
(1+0j)
```

2.7.4 Boolean: Bool()

This function is used to transform any data type into a bool type.

Example:

```
print(bool(0))
print(bool(1))
print(bool(15))
print(bool(15.6))
print(bool(0.33))
print(bool(0.0))
print(bool(15-6j))
print(bool(0+2.8j))
print(bool(0+0j))
print(bool("True"))
print(bool("False"))
print(bool(""))

False
True
True
True
True
False
True
True
False
True
True
False
```

2.7.5 String: Str()

This function is used to transform a number into a string or it is used to transform any data type to str data type.

Example:

```
print(str(15))
print(str(15.6))
print(str(15+6j))
print(str(True))

15
15.6
(15+6j)
True
```

All the basic data types are immutable, i.e. once the object is created, no modifications can be performed on them. If modifications are performed, then new objects will be created. The PVM won't let you create a new object, it will first verify whether any object with the same content is available or not, if it is available, then that object will be reused or a new object will be created. But the issue here is that several references point to the same object; by using one reference, if you are permitted to modify the content in the existing object then the rest of the references will be affected. To prevent this, the immutability concept is needed.

Example:

```
x= 15
y= 15
print(x is y)
print(id(x))
print(id(y))
```

```
True
140711611930720
140711611930720
```

2.7.6 Bytes Data Type: Bytes()

The bytes data type represents a group of byte numbers like an array. The only permitted values are from 0 to 256.

Example:

```
x1 = [3,6,8,10]
x2 = bytes(x1)
print(type(x2))
print(x2[0])
print(x2[-1])
for i in x2:
    print(i)
```

```
<class 'bytes'>
3
10
3
6
8
10
```

2.7.7 Byte Array Data Type: Bytearray()

This is similar to the bytes data type; besides that its elements can be altered.

Example:

```
x1 = [3,6,8,10]
x2 = bytearray(x1)
for i in x2:
    print(i)
x2[1]=20
print("After updates index value")
for i in x2:
    print(i)
```

```
3
6
8
10
After updates index value
3
20
8
10
```

2.7.8 List Data Type: List[]

The list type is used when you have to denote the combination of values as a single entity, which also allows insertion of duplicate values, heterogeneous objects, and is expandable in nature.

Example:

```
l=[11,11.6,'vijay',True,11]
print(l)
```

```
[11, 11.6, 'vijay', True, 11]
```

Example:

```
l=[11,22,33,44]
print(l[0])
print(l[-1])
print(l[1:4])
l[0]=300
for i in l:
    print(i)
```

```
11
44
[22, 33, 44]
300
22
33
44
```

A list is expandable in nature; you can increase or decrease the size of the list based on your requirement.

Example:

```
l1=[11,22,33,44]
l1.append("parth")
print(l1)
l1.remove(33)
print(l1)
l2=l1*2
print(l2)
```

```
[11, 22, 33, 44, 'parth']
[11, 22, 44, 'parth']
[11, 22, 44, 'parth', 11, 22, 44, 'parth']
```

2.7.9 Tuple Data Type: Tuple()

A tuple is immutable in nature, you cannot change its values once declared. They are declared using parenthesis.

Example:

```
tp=(11,22,32,45,54)
print(type(tp))
#tp[0]=600  #'tuple' object does not support item assignment'
#tp.append("swati") #'tuple' object has no attribute 'append'
#tp.remove(45)       #'tuple' object has no attribute 'remove'
print(tp)
```

```
<class 'tuple'>
(11, 22, 32, 45, 54)
```

2.7.10 Range Data Type: Range()

The range data type denotes a sequence of numbers. The elements declared inside the range data type are immutable in nature.

Case 1: range(3) print numbers from 0 to 2

Example:

```
x = range(3)
for i in x:
    print(i)

0
1
2
```

Case 2: range(1, 6) print numbers from 1 to 5

Example:

```
x = range(1,6)
for i in x:
    print(i)

1
2
3
4
5
```

Case3: range(1, 6, 2) Here, 2 means the increment value.

Example:

```
x = range(1,6,2)
for i in x:
    print(i)

1
3
5
```

Example: (to access elements using index)

```
x = range(1,6)
print(x[1])
print(x[8])

2

IndexError: range object index out of range
```

Range data type values can't be modified. The range data type does not allow values to be changed once declared.

Example: (creating list using range)

```
l1 = list(range(5))
print(l1)

[0, 1, 2, 3, 4]
```

2.7.11 Set Data Type: Set{}

The set data type is used when you have to denote the combination of values as a single entity. It does not permit insertion of duplicate values but permits mutable data, heterogeneous objects, and sets are expandable in nature, i.e. you can increase or decrease the size on the basis of your requirement.

The set elements cannot be accessed using an index.

Example:

```
x={60,0,11,34,11,'shashwat'}
print(x)
#print(x[0]) #'set' object is not subscriptable
x.add(80)
print(x)
x.remove(11)
print(x)

{0, 34, 11, 'shashwat', 60}
{0, 34, 11, 'shashwat', 80, 60}
{0, 34, 'shashwat', 80, 60}
```

2.7.12 Frozenset Data Type: Frozenset()

This is similar to the set data type except that it is immutable in nature; you cannot add or remove functions.

Example:

```
x={15,25,35,45}
y=frozenset(x)
print(type(y))
print(y)
for i in y:
    print(i)
#y.add(55)
#'frozenset' object has no attribute 'add'
#y.remove(35)
#'frozenset' object has no attribute 'remove'

<class 'frozenset'>
frozenset({25, 35, 45, 15})
25
35
45
15
```

2.7.13 Dictionary Data Type: Dict{}

The dictionary type is used when you want to denote a group of values as the key-value pairs; duplicate keys are not permitted but values can be permitted. If you wish to insert an entry with a duplicate key then the old value will be replaced with the new value.

Example:

```
d = {1:'vijay',2:'swati',3:'vimal'}
print(d)
d[1]='shashwat'
print(d)
#d.remove(1) #'dict' object has no attribute 'remove'

{1: 'vijay', 2: 'swati', 3: 'vimal'}
{1: 'shashwat', 2: 'swati', 3: 'vimal'}
```

To create empty dictionary:

$$p = \{\,\}$$

To add key-value pairs:

```
d={ }
d['a']='apple'
d['b']='banana'
d['c']='orange'
print(d)

{'a': 'apple', 'b': 'banana', 'c': 'orange'}
```

2.7.14 None Data Type: None

None says there is no value associated with it. It is used to fill the space when the value is not present.

Example:

```
def  m1():
    x=10
print(m1())

None
```

2.8 Escape Characters

In string literals, escape characters are preferred for performing specific tasks and their usage in code directs the compiler to perform a suitable action mapped to the adjacent character.

Example:

```
x="vijay\n mnnit"
print(x)
s="swati\t sharma"
print(s)
#s="vimal " kumar"
s="shashwat \" pathak"
print(s)
```

```
vijay
 mnnit
swati      sharma
shashwat " pathak
```

The various escape characters are mentioned below:

1. \n
2. \t
3. \r
4. \b
5. \f
6. \v
7. \'
8. \"
9. \\

and many more.

2.9 Input() Function

The input() function is used to accept an input from user. A programmer can ask a user to input a value by making the use of input(). It is used to read data directly, as required (Rossum 1995).

Example: (raw_input is used in Pyhon2; Python 3 doesn't support it)

```
i = raw_input("Enter input:")
print(type(i))
```

```
NameError: name 'raw_input' is not defined
```

In the input() function, every input value is considered as str type only.

Example:

```
i = input("Enter input:")
print(type(i))
j = input("Enter input:")
print(type(j))
k = input("Enter input:")
print(type(k))
l = input("Enter input:")
print(type(l))

Enter input:10
<class 'str'>
Enter input:vijay
<class 'str'>
Enter input:12.34
<class 'str'>
Enter input:False
<class 'str'>
```

Program: Write some Python code showing addition (+) and concatenation (+) using input() function.

```
s1=input("Enter S1:")
s2=input("Enter S2:")
s3 = int(s1)
s4 = int(s2)
print("Concatenation:",s1+s2)
print("Addition:",s3+s4)

Enter S1:15
Enter S2:40
Concatenation: 1540
Addition: 55
```

Program: Write some Python code to input two numbers and display their sum.

```
i=int(input("Enter the value-1:"))
j=int(input("Enter the value-2:"))
print("Addition:",i+j)

Enter the value-1:23
Enter the value-2:26
Addition: 49
```

Program: Write some Python code to read two values in a single line and show division operation on them.

```
s1,s2= [int(i) for i in input("Enter 2 input:").split()]
print("Division:", s1/s2)

Enter 2 input:12 6
Division: 2.0
```

Program: Write some Python code to read three float numbers with a comma (,) separator and display their multiplication.

```
s1,s2,s3= [float(i) for i in input("Enter 3 input :").split(',')]
print("Multiplication :",(s1*s2*s3))

Enter 3 input :2.1,3.1,4.1
Multiplication : 26.691
```

2.10 Evaluate: Eval() Function

Here, eval stands for evaluate. It takes a string as the parameter and returns it as if it is a Python expression (Van Rossum and Drake 2011).

Example:

```
e = eval("11+2+6")
print(e)
eval(input("Enter Expression"))

19
Enter Expression7+2*6/3

11.0
```

Example:

```
l1 = eval(input("List Input"))
print (type(l1))
print(l1)

List Input[2,4,6,7,9]
<class 'list'>
[2, 4, 6, 7, 9]
```

2.11 Command Line Arguments

Command line arguments are the arguments that are passed during run time. They are available in argv, which is available in the sys module.

Program: Write some code in command line argument.

```
from sys import argv
print("The Number of Command Line Arguments:", len(argv))
print("The List of Command Line Arguments:", argv)
print("Command Line Arguments one by one:")
for x in argv:
    print(x)
```

```
The Number of Command Line Arguments: 7
The List of Command Line Arguments: ['cli_program.py', '1', '2', '3', '4', '5', '6']
Command Line Arguments one by one:
cli_program.py
1
2
3
4
5
6
```

Example:

```
from sys import argv
sum=0
args=argv[1:]
for x in args:
    n=int(x)
    sum=sum+n
print("The Sum:",sum)
```

```
C:\Users\Lenovoi\Desktop\samarth>python cli_program.py 1 2 3 4 5 6
The Sum: 21
```

Example:

```
from sys import argv
print(argv[1])
```

```
Samarth
```

Example:

```
from sys import argv
print(argv[1]+argv[2])
print(int(argv[1])+int(argv[2]))
```

```
1020
30
```

2.12 Print: Print() Function

In Python, a function is a group of statements that are put together to perform a specific task. It is used to print the output (Lutz 2001).

Case 1: print() without any argument just it displays the new line character.

Case 2: Use of escape character, concatenation operator, repetition operator.

```
print("Vijay Sharma")
print("Swati \n Sharma")
print("Vimal\t Kumar")
print(3*"abc")
print("Python"*3)
print("Shashwat"+"Pathak")
```

```
Vijay Sharma
Swati
 Sharma
Vimal     Kumar
abcabcabc
PythonPythonPython
ShashwatPathak
```

Example:

```
print("Shashwat"+"Pathak")
print("OOPs","Programming")
```

```
ShashwatPathak
OOPs Programming
```

Case 3: print() with variable number of the arguments.

Example:

```
x,y,z=100,200,300
print("Data:",x,y,z)
```

```
Data: 100 200 300
```

By default, the output values are separated by space. If you wish to specify the separator use the "sep" attribute.

Example:

```
x,y,z=100,200,300
print(x,y,z,sep=',')
print(x,y,z,sep=':')
```

```
100,200,300
100:200:300
```

Case 4: print() with end attribute.

Example:

```
print("Hello")
print("vijay")
print("mnnit")
```

```
Hello
vijay
mnnit
```

If you want the output in the same line with space:

```
print("Hello",end=' ')
print("vijay",end=' ')
print("mnnit")
```

```
Hello vijay mnnit
```

Case 5: print(object) statement; you can pass any object as the argument to the print() statement.

Example:

```
l=[1,4,6,8]
t=(1,4,6,8)
print(l)
print(t)
```

```
[1, 4, 6, 8]
(1, 4, 6, 8)
```

Case 6: print() statement can be used with string and any number of arguments.

print(string, variable list).

Example:

```
s   ="vijay"
ids=100
s1 ="Book"
s2 ="Python"
print("Name",s,"Id-",a)
print(s1,"and",s2)
```

```
Name vijay Id- 6
Book and Python
```

Case 7: print (formatted string).

%i:	int
%f:	float
%d:	int
%s:	String type

Syntax: print("formatted string" %(variable list))

Program: Write some code in Python to display the values of two variables namely x and y.

```
x=20
y=30
print("%i"%x)
print("%d and %d"%(x,y))
```

```
20
20 and 30
```

Program: Write some code to enter the list of numbers and variables and display them.

```
x="vijay"
l=[10,20,30,40]
print("%s %s" %(s,1))
```

```
vijay [10, 20, 30, 40]
```

Case 8: print() with replacement operator {}.

Example:

```
n = "vijay"
ids = 2021
sub = "Python"
print("Name- {0} Id- {1} and Subject- {2}".format(n,ids,sub))
print("Name- {x} Id- {y} and Subject- {z}".format(x=n,y=ids,z=sub))
```

```
Name- vijay Id- 2021 and Subject- Python
Name- vijay Id- 2021 and Subject- Python
```

2.13 Delete Statement

For deleting in Python, the del keyword is used. After using a variable, it should be deleted by using the del keyword (Chun 2001).

Example:

```
a = 10
print(a)
del a
print(a)

10

NameError: name 'a' is not defined
```

After deleting a variable, you can't access that variable otherwise a NameError will be raised. In case of del, the variable will be removed, and you cannot access that variable. But in case of None, the variable will not be removed but the variable will be available for garbage collection.

Example:

```
x = "vijay"
print(x)
x = None
print(x)

vijay
None
```

2.14 Conclusion

After studying this chapter, the reader will be able to declare variables, choose data type, and perform base conversions and mathematical and logical operations. By studying all the concepts discussed in this chapter, the readers will agree with the fact that it is fun to write code in Python.

Review Questions

1. Explain character sets supported in Python.
2. Give the rules for naming an identifier in Python.
3. How are complex numbers displayed in interactive mode? Give an example.
4. Explain the use of the end keyword with a suitable example.
5. Explain escape characters.

Programming Assignments

PA 1: Write a program to read an integer as a string. Convert the string into an integer and display the type of value before and after converting to int.

PA 2: Write a program to calculate the perimeter of a rectangle.

PA 3: Write a program to display details entered by username, class, age, and gender.

PA 4: Write a program to enter the string from the user and convert it into upper case.

PA 5: Write a program to read and store the names of five states in different variables and print them.

References

Chun, Wesley. 2001. *Core python programming*. Vol. 1. Prentice Hall Professional.

Herscovics, Nicolas and Liora Linchevski. 1994. A cognitive gap between arithmetic and algebra. *Educational Studies in Mathematics* 27, no. 1: 59–78.

Kalicharan, Noel. 2015. *Learn to program with C*. Apress.

Lutz, Mark. 2001. *Programming Python*. O'Reilly Media.

Rossum, Guido. 1995. *Python reference manual*.

Van Rossum, Guido, and Fred L. Drake. 2011. *The python language reference manual*. Network Theory.

3

Operators in Python

3.1 Introduction

In the previous chapter, a brief overview of the math module was provided with examples. This chapter will help the reader to develop an understanding to perform arithmetic and logical operations in Python. To perform any mathematical operation, we define the operators. These operators are used in any programming language to define the mathematical and logical operations. Simple arithmetic and relational operators are used to define the mathematical operation and logical operators are used for Boolean and non-Boolean type behaviour with different definitions. Sometimes, we aim to perform bitwise operations, which can easily be dealt using a bitwise operator. Similarly, in different situations where we want to assign any particular value, assignment operator is used. Conditional, identity and membership operators are presented here to help readers understand designing different conditional conditions in their programs. Operator precedence has been presented to understand the order in which any operation will occur in multi-operator environment (Rai 2019). After reading this chapter, readers will be able to understand different operators and logical operands as well as their sequence of execution in Python. The presented subject matter along with the examples demonstrate the proper type and use of operators, which helps the reader to use an operator in different ways; this is required in many practical situations where code optimization is sought (Horton 2014).

3.2 Operators

The operator designates an operation that needs to be executed on data to produce the result. An operator requires data to operate, and this data is called an operand. Operators are the constructs that are used to change the value of operands (Campbell 2014). Python supports the below mentioned operators:

- Arithmetic operators
- Relational or comparison operators
- Logical operators
- Bit-wise operators

DOI: 10.1201/9781003185505-3

- Assignment operators
- Ternary or conditional operators
- Special operator.

3.2.1 Arithmetic Operators

These are used for performing basic arithmetic operations. The types of arithmetic operators are represented in Table 3.1.

TABLE 3.1
Arithmetic operators

Operations	Symbol
Addition	+
Subtraction	-
Multiplication	*
Division	/
Modulus	%
Floor division	//
Exponentiation	**

Example:

```
x=9
y=3
print('Addition=',x+y)
print('Subtraction=',x-y)
print('Multiplication=',x*y)
print('Division=',x/y)
print('Floor Division =',x//y)
print('Modulo=',x%y)
print('Exponent=',x**y)
```

```
Addition= 12
Subtraction= 6
Multiplication= 27
Division= 3.0
Floor Division = 3
Modulo= 0
Exponent= 729
```

When using the string concatenation operator (+), both the arguments should be of str type otherwise an error will be raised. When using the string multiplication operator (*), one argument has to be of int type and another of str type, otherwise an error will be raised.

Example:

```
print("vijay"+"30")
#print("vijay"+30)
#can only concatenate str (not "int") to str
print("Parth"*3)
#print("Parth"*"Sharma")
#can't multiply sequence by non-int of type 'str'
```

```
vijay30
ParthParthParth
```

3.2.2 Relational Operators

These compare the values of both sides and decide the relation between the values compared (Kalicharan 2008). The types of relational operators are represented in Table 3.2.

TABLE 3.2
Relational operators

Operations	Symbol
Less than equal to	<=
Less than	<
Greater than equal to	>=
Greater than	>

Program: Show the relational operations on integer values.

```
x=4
y=8
print(x>y)
print(x>=y)
print(x<y)
print(x<=y)

False
False
True
True
```

Program: Show all the relational operations on string values.

```
x="vimal"
y="vimal"
print(x>y)
print(x>=y)
print(x<y)
print(x<=y)

False
True
False
True
```

Example:

```
print(True>True)
print(True>=True)
print(15 >True)
print(False > True)
#print(15>'swati')
#not supported between instances of 'int' and 'str'

False
True
True
False
```

Example:

```
x=3
y=8
if(x>y):
    print("x")
else:
    print("y")

y
```

Python allows a chaining comparison of relational operators; if all the comparisons are True then only True will be returned, whereas if any single comparison is not leading as True, the overall output will be False.

Example:

```
print(2<12)
print(12<25<35)
print(12<27<38<45)
print(15<22<29<42>58)

True
True
True
False
```

Equality operators: ==,!=

These operators can be performed over any data type and it is also applicable for incompatible type and chaining concept (Chivers and Sleightholme 2018).

Example:

```
print(12==23)
print(12!= 23)
print(12==True)
print(False==False)
print("vijay"=="vijay")
print(15=="vijay")

False
True
False
True
True
False
```

Example:

```
print(1==4==6==9)
print(4==4==4==4)

False
True
```

3.2.3 Logical Operators

The types of logical operators are represented in Table 3.3

TABLE 3.3

Logical operators

Operations	Symbol
And	∧
Or	∨
Not	~

They are used as follows:

- **For Boolean type behaviour:**

and: The result will be True if both the arguments are True.
or: The result will be True if any one of the argument is True.
not: The result will be a complement of the argument.

Example:

```
print(True and False)
print(not False)
print(False or True)
print(True and True)

False
True
True
True
```

- **For non-Boolean types of behaviour:**

Here, 0 and an empty string denotes False whereas non-zero denotes True.

- x and y:

If the first part is evaluated as False, then False will be returned otherwise the second part will be returned.

Example:

```
print(1 and 2)
print(0 and 2)

2
0
```

- x or y:

If the first part is evaluated as True, then True will be returned otherwise the second part will be returned.

Example:

```
print(1 or 2)
print(0 or 2)

1
2
```

- not x:

The result will be True if x evaluates to False and vice-versa.

Example:

```
print(not 11)
print(not 0)

False
True
```

Example:

```
print("vijay" and "vijaymnnit")
print(""    and "vijay")
print("" or "vijay")
print("vijay" or "")
print(not  "")
print("vijay" and "")
print(not "vijay")

vijaymnnit

vijay
vijay
True

False
```

3.2.4 Bitwise Operators

These deal with bits and execute an operation bit-by-bit. The types of bitwise operators are shown in Table 3.4

TABLE 3.4

Bitwise operators

Symbol	Operations
|	Bitwise OR operator
&	Bitwise AND operator
^	Bitwise NOT operator
~	Bitwise complement operator
<<	Bitwise left operator
>>	Bitwise right operator

These operators can be applied bit-wise. They are relevant for int and Boolean data types and give errors for the rest of the data types.

Bitwise AND Operator (&)

If both the arguments are 1 then only the result will be 1; otherwise 0.

Bitwise OR Operator (|)

If at least one of the bits is 1 then only the result will be 1; otherwise 0.

Bitwise NOT Operator (^)

The result will be 1 if both the bits are distinct otherwise the result will be 0.

Example:

```
print(6&8)
print(4|5)
print(6^8)
```

```
0
5
14
```

Bit-wise Complement Operator (~)

If the bit designates to 1, its complement will be 0 whereas if the bit designates to 0, its complement will be 1 (Yalag, Trupti, and Nirgude 2016).

Example:

```
print(~8)
```

```
-9
```

Bitwise Left Shift Operator (<<)

After shifting the bits to the left, fill the empty cells with 0.

Example:

```
print(10<<3)
```

```
80
```

Bitwise Right Shift Operator (>>)

After shifting the bits to the right, fill the empty cells with the signed bit.

```
print(10>>3)
```

```
1
```

They are pertinent for Boolean types also.

Example:

```
print(True & False)
print(True | False)
print(True ^ False)
print(~ False)
print(True<<2)
print(False<<3)
```

```
False
True
True
-1
4
0
```

3.2.5 Assignment Operators

The assignment operator assigns values to the operand. The assignment operators are shown in Table 3.5

TABLE 3.5

Assignment operators

Operators
+=
-=
*=
/=
%=
//=
**=
&=
|=
^=
>>=
<<=

They are used for assigning the values to the variable. They can be merged with a different operator to produce a compound assignment operator.

Example:

```
a=15
a+=25
print(a)
```

```
40
```

Example:

```
a=15
a&=6
print(a)
```

```
6
```

3.2.6 Ternary Operator or Conditional Operator

If the condition designates to True then first_value will be regarded otherwise second_value will be regarded.

Syntax: p = first_value if condition else second_value

Example:

```
x,y=20,30
x=40 if x<y else 50
print(x)
```

```
40
```

Program: Using a ternary operator, write a program to read two numbers and print the minimum number.

```
x=int(input("Enter the X:"))
y=int(input("Enter the Y:"))
min=x if x<y else y
print("Min:",min)
```

```
Enter the X:20
Enter the Y:30
Min: 20
```

Program: Using a ternary operator, write a program to read three numbers and print the minimum number.

```
x=int(input("Enter the X:"))
y=int(input("Enter the Y:"))
z=int(input("Enter the Z:"))
min=x if x<y and x<z else y if y<z else z
print("Min:",min)
```

```
Enter the X:8
Enter the Y:4
Enter the Z:6
Min: 4
```

Program: Using a ternary operator, write a program to read three numbers and print the maximum number.

```
x=int(input("Enter the X:"))
y=int(input("Enter the Y:"))
z=int(input("Enter the Z:"))
max=x if x>y and x>z else y if y>z else z
print("Max:",max)
```

```
Enter the X:5
Enter the Y:18
Enter the Z:9
Max: 18
```

3.2.7 Special Operators

The types of special operators are shown in Table 3.6

TABLE 3.6
Special operators

Operators
Identity operator
Membership operator

3.2.7.1 Identity Operator

Identity operators contrast the memory location of two objects. They are inherited for address comparison also. There are two identity operators: is and is not.

- p1 is p2   produces True if p1 and p2 are indicating to the same object.
- p1 is not p2 produces True if p1 and p2 are not indicating to the same object.

Example:

```
x=14
y=18
print(x is y)
x1=True
y1=True
print( x1 is y1)
```

```
False
True
```

Example:

```
x="vijay"
y="vijay"
print(id(x))
print(id(y))
print(x is y)
```

```
2644860727152
2644860727152
True
```

Example: (for comparing the content, == operator is used)

```
l1=["vijay","vimal","swati"]
l2=["vijay","vimal","swati"]
print(id(l1))
print(id(l2))
print(l1 is l2)
print(l1 is not l2)
print(l1 == l2)
```

```
2644860401408
2644859447488
False
True
True
```

3.2.7.2 Membership Operator

These are used to test whether the given object is available in the prescribed collection or not.

- in   produces True if the given object is available in the prescribed collection
- not in produces True if the given object is not available in the given collection

Example:

```
str1="Teaching is very difficult task"
print('a' in str1)
print('v' in str1)
print('d' not in str1)
print('task' in str1)

True
True
False
True
```

Example:

```
l1=["vijay","vimal","swati"]
print("vijay" in l1)
print("vimal" in l1)
print("swati" not in l1)

True
True
False
```

3.3 Operator Precedence

Table 3.7 shows the precedence of all the operators. Operator associativity identifies the order of the evaluation in case of the same precedence, and is not grouped by the parenthesis. An operator may be left or right associative. In left-associative, the operators on the left side will be evaluated first whereas in right-associative, operators on the right side will be evaluated first (Qamar and Raza 2020).

TABLE 3.7

Operator precedence

Operator	Description
**	Exponentiation
-,+,-	Complement, unary plus and minus
*,/,%,//	Multiply, divide, modulo and floor division
+,-	Addition and subtraction
>>,<<	Right and left bit-wise shift
&	Bit-wise AND
^, |	Bit-wise exclusive OR
<=,<>, >=	Comparison operator
<>, ==,!=	Equality operator
=,%=,/=,//=,-=,+=,*=,**=	Assignment operator
is, is not	Identity operator
in, not in	Membership operator
not, or, and	Logical operator

Example:

```
print(4+15*3)
print((4+15)*3)
```

```
49
57
```

Example:

```
p=5
q=7
r=3
s=2
print((p+q)*r/s)
print((p+q)*(r/s))
print(p+(q*r)/s)
```

```
18.0
18.0
15.5
```

Program: Write a program to enter principal, rate of interest, and time in years from user and calculate simple interest.

```
P = eval(input('Enter the Principal Value:'))
R = eval(input('Enter the Rate Value:'))
T = eval(input('Enter the Time Value:'))
SI= (P*R*T)/100
print('The simple interest is:',SI)
```

```
Enter the Principal Value:1000
Enter the Rate Value:4
Enter the Time Value:2
The simple interest is: 80.0
```

Program: Write a program to read temperature in Celsius from user and convert it into Fahrenheit.

```
P = eval(input("Enter temperature in celsius"))
F = (9/5)*P
print("Fahrenheit:",F)
```

```
Enter temperature in celsius98
Fahrenheit: 176.4
```

3.4 Conclusion

This chapter presented information regarding use, type, and applicability of different operators in Python. Different use cases, conditions, and application of varied class of arithmetic and logical operators for simple and Boolean operands were presented with illustrations. A clear understanding of these functions is a primary skill to start programming in Python. The practice problems given at the end of this chapter will help the reader to start with basic programming in Python. The next chapter will help develop the understanding of control flow, which are the internal instructions flowing into the compiler to execute the operation.

Review Questions

1. Explain the use of the // operator.
2. Explain the use of the ** operator.
3. Explain the use of the is operator.
4. Briefly explain the various types of operators in Python.
5. List the operator precedence in Python.

Programming Assignments

PA 1: Write some Python code to take the principal, rate of interest, and number of years from the user and calculate the compound interest.

PA 2: Write some Python code to reverse a five digit number using % and //.

PA 3: Write some Python code to shift input data by 5 bits towards the left.

PA 4: Write some Python code to read a five digit number through the keyboard and calculate the sum of its digits.

PA 5: Write some Python code to read the marks of five subjects and calculate the sum and percentage of the student.

References

Campbell, Matthew. 2014. *Objective-C quick syntax reference*. Apress.

Chivers, Ian D., and Jane Sleightholme. 2018. *Introduction to programming with Fortran*. Vol. 2. Springer.

Horton, Ivor. 2014. *Beginning C++*. Apress.

Kalicharan, Noel. 2008. *C programming – a beginner's course*. Create Space.

Qamar, Usman and Muhammad Summair Raza. 2020. Data science programming languages. In *Data Science Concepts and Techniques with Applications*. Springer, Singapore: 153–196.

Rai, Laxmisha, ed. 2019. *Programming in C++: Object oriented features*. Vol. 5. Walter de Gruyter GmbH & Co KG.

Stine, James E. 2004. *Digital computer arithmetic data path design using verilog HDL*. Springer Science & Business Media.

Yalagi, Pratibha S., Trupti S. Indi, and Manisha A. Nirgude. 2016. Enhancing the cognitive level of novice learners using effective program writing skills. International Conference on Learning and Teaching in Computing and Engineering (LaTICE). IEEE: 167–171.

4

Control Flow in Python

4.1 Introduction

A control statement identifies the controlled circulation of set-of-instructions in which they will execute. A controlled statement can comprise one or more instructions. There are three basic methods of control movement in it, i.e. sequential flow, selection flow, and iterative flow. The code is always executed from the first to the last line of the program; known as sequential control flow (De Brouwer 2018). However, in some cases you want to process only a few set-of-statements or execute set-of-statements again and again. Thus, the decision control statements can alter the flow of a sequence of instructions. Such types of conditional processing provided by the decision control statements extend the usefulness of programs. It allows the programmers to build a program that determines which statements of the code should be executed and which should be ignored. Figure 4.1 shows the categorization of decision control statements.

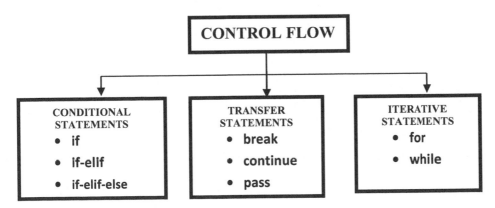

FIGURE 4.1
Categorization of decision control statements

4.2 Conditional Statements

Python supports various decision-making statements. These are discussed as follows:

DOI: 10.1201/9781003185505-4

4.2.1 If Statement

The if statement executes the statement if the given condition is True.

Syntax:
  if condition: statement

  OR.

  if condition:
  statement1
  statement2
  statement3

Example:

```
n=input("Enter Name:")
if n=="vijay" :
    print("Correct Information")
print("Done")
```

```
Enter Name:vv
Done
```

4.2.2 If-Else Statement

The if-else statement takes care of a True as well as False condition. If the condition given in if is True, then Action1 will be processed, otherwise Action2 will be processed.

if condition:
        Action1
else:
        Action2

Example:

```
n=input("Enter Name:")
if n=="vimal":
    print("Correct Information")
else:
    print("Not Correct Information")
```

```
Enter Name:vim
Not Correct Information
```

```
Enter Name:vimal
Correct Information
```

4.2.3 If-Elif-Else Statement

The if-elif-else statement takes care of multiple conditions. If the condition given in if is True, then Action1 will be processed, if the condition given as condition2 is True, then Action2 will be processed, if all the elif conditions are False, then the default action given in else will be processed.

Syntax:

> if condition:
>> Action1.
> elif condition2:
>> Action2
> elif condition3:
>> Action3
> elif condition4:
>> Action4
> else:
>> default action

Example:

```
n=input("Enter Input:")
if n=="Man" :
    print("This is Male ")
elif n=="Women":
    print("This is Female")
elif n=="Kids":
    print("This is child")
else :
    print("No Information")
```

```
Enter Input:Women
This is Female
```

Program: Write some code to identify the largest of three numbers from the command prompt.

```
n1=int(input("Enter N1:"))
n2=int(input("Enter N2:"))
n3=int(input("Enter N3:"))
if n1>n2 and n1>n3:
    print("Greatest Value:",n1)
elif n2>n3:
    print("Greatest Value:",n2)
else :
    print("Greates Value:",n3)
```

```
Enter N1:23
Enter N2:12
Enter N3:47
Greates Value: 47
```

Program: Write some Python code to check that the number lies between 1 and 10 or not.

```
n=int(input("Enter Input:"))
if n>=1 and n<=10 :
    print("Under 10:",n)
else:
    print("Out of 10:",n)
```

```
Enter Input:40
Out of 10: 40

Enter Input:9
Under 10: 9
```

Program: Write some Python code that displays days of the week.

```
n=int(input("Enter Input:"))
if n==0 :
    print("Sunday")
elif n==1:
    print("Monday")
elif n==2:
    print("Tuesday")
elif n==3:
    print("Wednesday")
elif n==4:
    print("Thrusday")
elif n==5:
    print("Friday")
elif n==6:
    print("Saturday")
else:
    print("No Choice")
```

```
Enter Input:5
Friday
```

4.3 Iterative Statements

If you want to process the group of statements a number of times then you should favor iterative statements. Python provides two kinds of iterative statements.

4.3.1 For Loop

The for loop in Python is slightly different from other programming languages. In Python, the for loop iterates through a sequence of objects, i.e. it iterates through each value in a sequence. If you wish to process something on every element available in some series then you should select the for loop (Lutz 2001).

Syntax: for p in sequence:
                    body

Where, the sequence can be any string or collection.
              body is executed for all the elements available in sequence.

Program: Write some Python code that displays the characters available in the given string.

```
n="Swati"
for i in n:
    print(i)
```

```
S
w
a
t
i
```

Program: Write some Python code that displays the characters available in the string index-wise.

```
n=input("Enter some String: ")
i=0
for n1 in n:
    print(i,"index:",n1)
    i=i+1

Enter some String: vijay
0 index: v
1 index: i
2 index: j
3 index: a
4 index: y
```

Program: Write some Python code that displays 'vimal' three times in different lines, using a for loop.

```
for i in range(3) :
    print("vimal")

vimal
vimal
vimal
```

Program: Using the range function, print the numbers from 0 to 4.

```
for i in range(4):
    print(i)

0
1
2
3
```

Program: Using the range function, display odd numbers from 0 to 8.

```
for i in range(8) :
    if(i%2!=0):
        print(i)

1
3
5
7
```

Program: Using the range function, display numbers from 3 to 1 in decreasing order.

```
for i in range(3,0,-1):
    print(i)

3
2
1
```

Program: Write a program to enter a list from the user and print the sum of its elements.

```
l = eval(input("Enter List:"))
s=0
for i in l:
    s=s+i
print("Addition=",s)
```

```
Enter List:[3,5,7,9]
Addition= 24
```

4.3.2 While Loop

The while loop is a loop control statement in Python and frequently used in programming for repeated execution of the statements in a loop (Sanner 1999). It processes the sequence of statements until the condition will be True. If you want to process a set-of-statements sequentially until some condition is False, then you should choose a while loop.

Syntax: while condition:
                  body

Program: Write some code to display numbers 1 to 4 using a while loop.

```
a = 1
while a <= 4:
    print(a)
    a = a+1
```

```
1
2
3
4
```

Program: Write some code to print the addition of the first n numbers.

```
x=int(input("Enter Input:"))
s=0
i=1
while i<=x:
    s=s+i
    i=i+1
print("Addition" ,x, ":",s)
```

```
Enter Input:6
Addition 6 : 21
```

Program: Write a program to prompt the user to enter the names until entering the name 'Vijay'.

```
n=""
while n!="vijay":
    n=input("Enter Input:")
print("Done")
```

```
Enter Input:vijay
Done
```

Program: Write a program showing the use of an infinite loop.

```
i = 0;
while True:
    i = i+1;
    print("Python",i)
```

```
Python 1
Python 2
Python 3
Python 4
Python 5
Python 6
Python 7
Python 8
Python 9
```

4.3.3 Nested Loops

The for and while loop statements can be nested in the same manner in which the if statements are nested. Loops within the loops or when one loop is inserted completely within another loop are called nested loops (Zelle 2004).

Example:

```
for i in range(3):
    for j in range(3):
        print("i=",i," j=",j)
```

```
i= 0    j= 0
i= 0    j= 1
i= 0    j= 2
i= 1    j= 0
i= 1    j= 1
i= 1    j= 2
i= 2    j= 0
i= 2    j= 1
i= 2    j= 2
```

4.4 Transfer Statements

4.4.1 Break Statement

The keyword break permits the programmer to terminate the loop. The loop gets immediately terminated on encountering the break statement in the loop and the control of the program goes to the first statement, following the loop. You can choose a break statement inside the loop to break the loop execution on the basis of a constraint.

In the following program, four elements are taken in a list. If the values of elements are greater than 18, then the print statement will be printed and the break statement will take hold, but if the value is smaller than 18, then directly the value will be printed.

Example:

```
l=[15,18,25,45]
for i in l:
    if i>18:
        print("This is the location")
        break
    print(i)
```

```
15
18
This is the location
```

Example:

```
for i in range(8):
    if i==4:
        print("Stop here...")
        break
    print(i)
```

```
0
1
2
3
Stop here...
```

4.4.2 Continue Statement

A continue statement is the exact opposite of a break statement. On encountering the continue statement inside the loop, the rest of the statements inside the body will be skipped and the loop condition will be verified to check whether the loop should continue or should terminate. A continue statement can be used to skip the on-going iteration and continues with the next iteration.

Example:

```
l=[15,17,25,45,65,9,12]
for i in l:
    if i>18:
        continue
    print(i)
```

```
15
17
9
12
```

Example:

```
n=[11,25,5,30]
for n1 in n:
    if n1==0:
        continue
    print("30/{} = {}".format(n1,30/n1))
```

```
30/11 = 2.727272727272727
30/25 = 1.2
30/5 = 6.0
30/30 = 1.0
```

Example:

```
l=[11,25,35,45]
for i in l:
    if i>=60:
        break
    print(i)
else:
    print("Done")

11
25
35
45
Done
```

Program: Write a program to display the odd numbers in the range 0 to 8.

```
for i in range(8):
    if i%2==0:
        continue
    print(i)

1
3
5
7
```

4.4.3 Pass Statement

Depending on the program used, if a block is needed that will not execute anything then, you can declare the empty block using a pass keyword: a keyword in Python.

Example:

```
for i in range(8):
    if i%2==0:
        print(i)
    else:
        pass

0
2
4
6
```

4.5 Loops with Else Block

When processing a loop, if a break statement is not processed, then only the else part will be processed; it refers to a loop without the break (Learner and Powell 1974).

Example:

```
l=[15,26,38,49,60]
for i in l:
    if i>=200:
        print("Not Processing")
        break
    print(i)
else:
    print("Done")
```

```
15
26
38
49
60
Done
```

Example:

```
l=[15,25,300,400,29,38,60]
for i in l:
    if i>=200:
        print("Not Processing")
        break
    print(i)
else:
    print("Done")
```

```
15
```

4.6 Conclusion

In this chapter, we learned about the available control flow statements in the Python programming language named if, else, elseif, for, while, break, and continue with sufficient examples.

Review Questions

1. Explain various decision-making statements in Python.
2. What is the purpose of a break statement.
3. Explain the usage of a continue statement.
4. Explain a for loop with else with an example.
5. Explain different types of loops with examples.

Programming Assignments

PA 1: Write a program that prints the sum of every tenth number from 0 to 100, using a while loop.

PA 2: Write some Python code to check whether the entered number is prime or not, using a break keyword.

PA 3: Write some Python code to find the sum of the digits of a given number.

PA 4: Write a program that asks the user to enter the day of the week; if the entered day of the week is between 1 and 7, then display the respective name of the day.

PA 5: Write a program to identify whether the entered year is a leap year or not.

References

De Brouwer, Philippe J.S. 2018. *A practical introduction to quantitative methods and R*. No publisher.

Langtangen, Hans Petter. 2012. *A primer on scientific programming with Python*. Springer Berlin Heidelberg.

Learner, Arnold, and Anthony J. Powell. 1974. *An introduction to ALGOL 68 through problems*. Macmillan.

Lutz, Mark. 2001. *Programming Python*. O'Reilly Media.

Sanner, Michel F. 1999. Python: A programming language for software integration and development. *J Mol Graph Model* 17, no. 1: 57–61.

Zelle, John M. 2004. *Python programming: An introduction to computer science*. Franklin, Beedle & Associates.

5

Strings

5.1 Introduction

In Python, a string is an object of the str class. This string class has many constructors. Its sequence is made up of one or more individual characters, where a character could be a letter, digit, white space, or any other symbol (Walters 2014). Python treats strings as contiguous series of characters delimited by single, double, or even triple quotes.

Syntax: s = 'vijay'   or   s = "vijay"

Example:

```
ch = 'a'
type(ch)

str
```

5.2 Multiline String Literals

The multiline string literals are declared using triple single quotes or double quotes.

Example: p = '''vijaymnnitallahabad'''

Triple quotes can also be used for using single or double quotes as symbols in string literal.

DOI: 10.1201/9781003185505-5

Example:

```
#p = 'This is 'single quote symbol'
q = 'This is \' single quote symbol'
r = "This is ' single quote symbol"
s = 'This is " double quotes symbol'
#t = 'The "Python Notes" by 'vijay' is very helpful'
#u = "The "Python Notes" by 'vijay' is very helpful"
v = 'The \"Python Notes\" by \'vijay\' is very helpful'
w = '''The "Python Notes" by 'vijay' is very helpful'''
print(q)
print(r)
print(s)
print(v)
print(w)
```

```
This is ' single quote symbol
This is ' single quote symbol
This is " double quotes symbol
The "Python Notes" by 'vijay' is very helpful
The "Python Notes" by 'vijay' is very helpful
```

5.3 Accessing Characters of String

The characters of the string can be accessed by the following procedures:

5.3.1 By Using Index

Python supports a positive as well as negative index. The positive index designates a left-to-right direction whereas the negative index designates a right-to-left direction (Celko 2010).

Example:

```
a='vijay'
print(a[0])
print(a[4])
print(a[-1])
#print(a[10])   #dexError: string index out of range
```

```
v
y
y
```

Program: Write some Python code to input a string and print the characters according to a positive index and negative index.

```
n=input("Enter input:")
i=0
for s in n:
    print("Positive index {} Negative index {}  is {}"
          .format(i,i-len(n),s))
    i=i+1
```

```
Enter input:vijay
Positive index 0 Negative index -5  is v
Positive index 1 Negative index -4  is i
Positive index 2 Negative index -3  is j
Positive index 3 Negative index -2  is a
Positive index 4 Negative index -1  is y
```

5.3.2 By Using Slice Operator

Syntax: p[begin_index: end_index: step_value]
   where, begin_index is the position from where to consider the slice
          end_index is the position from where to end the slice
          step_value is the augmented value

Example:

```
s="My Father is best teacher"
print(s[1:7:1])
print(s[1:7])
print(s[1:7:2])
print(s[:7])
print(s[7:])
print(s[::])
print(s[:])
print(s[::-1])
```

```
y Fath
y Fath
yFt
My Fath
er is best teacher
My Father is best teacher
My Father is best teacher
rehcaet tseb si rehtaF yM
```

5.3.2.1 *Behavior of Slice Operator*

1. p [begin_index: end_index: step_value]
2. The step_value can be positive or negative
3. If positive, then it follows forward (ahead) direction
4. If negative, then it should be backward (reverse) direction.

In forward direction:

The default .value for begin is 0
The by default for end is the length of string
The by default .for step is incremented by one.

In backward direction:

The by default for begin is decremented by one.
The by-default for end is [- (string_length. + 1)]

5.3.2.2 *Slice Operator Case Study*

Example:

```
s = 'abcdefghij'
print(s[1:6:2])
print(s[5:0:1])
print(s[::1])
print(s[::-1])
print(s[3:7:-1])
print(s[7:4:-1])
print(s[0:10000:1])
print(s[0:-9:-2])
print(s[-4:1:-1])
print(s[-4:1:-2])
print(s[0:-10:-1])
print(s[0:-11:-1])
print(s[0:0:1])
print(s[-5:-9:-2])
print(s[10:-1:-1])
print(s[10000:2:-1])
#s[9:0:0]  # ValueError: slice step cannot be zero
```

```
bdf

abcdefghij
jihgfedcba

hgf
abcdefghij

gfedc
gec

a

fd

jihgfed
```

5.4 Mathematical Operators for Strings

The following mathematical operators can be applied on strings:

1. Concatenation operator (+)
2. Repetition operator (*)

Example:

```
print("vijay"+"mnnit")
print("vijay"*2)

vijaymnnit
vijayvijay
```

5.5 Len() Function

The len() function is preferred for finding the number of characters present in a string (Stine 2004).

Example:

```
n = 'vijay'
print(len(n))

5
```

Program: Write a program using a while loop to access all the characters of the string in a forward direction and backward direction.

```
a = "Ankur is a Phd Scholor"
n = len(a)
i = 0
print("Forward direction")
while i<n:
    print(a[i],end=' ')
    i +=1
print("\nBackward direction")
i = -1
while i >= -n:
    print(a[i],end=' ')
    i = i-1

Forward direction
A n k u r   i s   a   P h d   S c h o l o r
Backward direction
r o l o h c S   d h P   a   s i   r u k n A
```

Alternative way

```
a = "Ankur is a Phd Scholor"
print("Original String")
for i in a:
    print(i,end=' ')
print("\nForward direction")
for i in a[::]:
    print(i,end=' ')
print("\nBackward direction")
for i in a[::-1]:
    print(i,end=' ')
```

```
Original String
A n k u r   i s   a   P h d   S c h o l o r
Forward direction
A n k u r   i s   a   P h d   S c h o l o r
Backward direction
r o l o h c S   d h P   a   s i   r u k n A
```

5.6 Checking Membership

The membership operators can be used with strings to determine whether a string is present in another string. Therefore, 'in' and 'not in' operators are also known as membership operators (Scholtz and Wiedenbeck1990).

Program: Write a Python code to find whether the sub-string is found in main string.

```
str1 = 'vijay'
print('j' in str1)
print('w' in str1)
str1 = input("Enter String-1:")
substring = input("Enter Sub-String:")
if substring in str1:
    print(substring,"Matched Sting-1")
else:
    print(substring,"Not Matched Sting-1")
```

```
True
False
Enter String-1:Vijay is a book author
Enter Sub-String:Vijay
Vijay Matched Sting-1

Enter Sub-String:Parth
Parth Not Matched Sting-1
```

5.7 String Operations

5.7.1 Comparison of Strings

Operators such as ==, <, >, <=,>=,!= are preferred for comparing the strings. Python contrasts the strings by comparing their corresponding characters.

Program: Write a program to compare two strings in alphabetical order.

```
s1=input("Enter String-1:")
s2=input("Enter String-2:")
if s1==s2:
    print("Equal String-1 and String-2")
elif s1<s2:
    print("String-1 is smallest of String-2")
else:
    print("String-1 is bigger of String-2")
```

```
Enter String-1:Vijay
Enter String-2:Ajay
String-1 is bigger of String-2

Enter String-1:Shashwat
Enter String-2:Vimal
String-1 is smallest of String-2
```

5.7.2 Removing Spaces from a String

Python provides various methods to remove white space characters from the beginning, end, or both the ends of a string.

1. **rstrip():** It returns the output with the trailing white space characters eliminated.
2. **lstrip():** It returns the output with the leading white space characters eliminated.
3. **strip():** It returns the output with the leading and trailing white space characters eliminated.

Example:

```
s=input("Enter the String:")
m=s.strip()
if m=='India':
    print("Welcome to India")
elif m=='England':
    print("Welcome to England")
elif m=="America":
    print("Welcome to America")
else:
    print("Sorry Sir")
```

```
Enter the String:ind
Sorry Sir

Enter the String:India
Welcome to India
```

5.7.3 Finding Sub-strings

In Python, the following methods are used for finding sub-strings:

- In the case of forward direction; find() and index() are used.
- In the case of backward direction; rfind() and rindex() are used.

find(): It gives the index of the first occurrence where the string str1 is found in the main string or returns −1 if the string str1 is not found in this string.

Syntax: s.find (.substring)

Example:

```
p="Paragraph on importance of good reading habits"
print(p.find("good"))
print(p.find("on"))
print(p.find("Paragraph"))
print(p.rfind("w"))
```

```
27
10
0
-1
```

Example:

```
p="Paragraph on importance of good reading habits"
print(p.find('i'))
print(p.find('i',7,15))
print(p.find('b',7,15))
```

```
13
13
-1
```

5.7.4 Index()

This is quite similar to the find() method, the only difference is if the mentioned sub-string is not available, then the ValueError will be raised.

Program: Write a program to identify whether the sub-string is found in the main string.

```
str1=input("Enter Original String:")
substring=input("Enter SubString:")
try:
    n=str1.index(substring)
except ValueError:
    print("SubString not matched")
else:
    print("SubString matched")
```

```
Enter Original String:vijay
Enter SubString:vijay
SubString matched
```

```
Enter Original String:vijay
Enter SubString:vimal
SubString not matched
```

Program: Write a program to print all the positions of the sub-string from the main string.

```
str1=input("Enter Original String:")
substring=input("Enter SubString:")
flag=False
position=-1
n=len(str1)
while True:
    position=str1.find(substring,position+1,n)
    if position==-1:
        break
    print("Position is",position)
    flag=True
if flag==False:
    print("Sorry...Not a Position")
```

```
Enter Original String:aabddfgrhystt
Enter SubString:w
Sorry...Not a Position
```

```
Enter Original String:pqppxyqqyy
Enter SubString:y
Position is 5
Position is 8
Position is 9
```

5.7.5 Methods to Count Sub-string in the Main String

The count() method is used to identify the number of occurrences of the sub-string in the main string.

1. s.count (sub_string): It will find the sub_string in the entire string.
2. s.count(sub_string: begin_index: end_index): It searches from the beginning to the end of the index (Huff 2006).

Example:

```
s="ppabjcdbjbhgetyubghhjyyttrr"
print(s.count('y'))
print(s.count('hh'))
print(s.count('j',2,9))

3
1
2
```

5.7.6 Replacing a String with Another String

It returns a new string that replaces all occurrences of the previous string with a new string.

s.replace (previousstring, newstring)

Example:

```
str1 = "Teaching is a very tough task"
str2 = str1.replace("tough","easy")
print(str2)
```

```
Teaching is a very easy task
```

Example: (all occurrences of 'p' will be replaced by 'q')

```
str1 = "pqpqpqpqpqpqpqpqppqppq"
str2 = str1.replace("p","q")
print(str2)
```

```
qqqqqqqqqqqqqqqqqqqqqqq
```

Modification of contents of string object using replace() method:

The string is immutable in nature, but one can change the contents by using the replace() method. But using the replace() method to change the contents of a string will produce a new string. No changes will be performed in the existing string.

Example:

```
str1 = "pqpqppqpqq"
str2 = str1.replace("p","q")
print(str1,"address is :",id(str1))
print(str2,"address is :",id(str2))
```

```
pqpqppqpqq address is : 2000412255920
qqqqqqqqqq address is : 2000412239792
```

5.7.7 Splitting of Strings

On using the split() method, it produces a series of distinct words present in the string or it is used to break up a string into smaller strings. The by-default separator is space; its return type is list.

Syntax: p = s.split(separator)

Example:

```
s = "Parth is a good boy"
p = s.split()
for i in p:
    print(i)
```

```
Parth
is
a
good
boy
```

Example:

```
d="15-01-2021"
l=d.split('-')
for x in l:
    print(x)
```

```
15
01
2021
```

5.7.8 Method to Join the Strings

A group of strings can be joined with respect to the given separator.

    s = separator.join(group of strings)

Example:

```
t = ('vijay', 'vimal', 'swati')
a = '-'.join(t)
print(a)
```

```
vijay-vimal-swati
```

Example:

```
l = ['Vijay', 'Jaunpur', 'Parth', 'Delhi']
a = ':'.join(l)
print(a)
```

```
Vijay:Jaunpur:Parth:Delhi
```

5.7.9 Methods to Change the Case of a String

A string may be present in lower case or upper case. The string in lower case can be converted into upper case and vice versa using various methods of the str class (Yalag, Trupti, and Nirgude 2016).

1. upper(): It produces the string with the letters converted into upper case.
2. lower(): It produces the string with the letters converted into lower case.

3. swapcase(): It produces the string that converts upper case characters into lower case characters and vice versa.
4. title(): It produces the string with the first letter capitalized in all the words of the string.
5. capitalize(): It produces the string with only the first character capitalized.

Program: Write a program showing different methods to change the case of a string.

```
s = 'Teaching is a good job'
print(s.upper())
print(s.lower())
print(s.swapcase())
print(s.title())
print(s.capitalize())
```

```
TEACHING IS A GOOD JOB
teaching is a good job
tEACHING IS A GOOD JOB
Teaching Is A Good Job
Teaching is a good job
```

5.7.10 Method to Check the Start and End Part of String

Python possesses the below mentioned methods for checking the starting and ending part of the string:

- s.startswith(substring)
- s.endswith(substring)

Example:

```
s = 'Teaching is a good job'
print(s.startswith('job'))
print(s.endswith('Teaching'))
print(s.endswith('job'))
```

```
False
False
True
```

5.7.11 Methods for Checking the Type of Characters

A string may contain digits, alphabets, or a combination of both. Thus, various methods are available to test if the entered string is a digit or alphabet or is alphanumeric (Chivers and Sleightholme 2018).

1. isalnum(): It produces True if characters in the given string are alphanumeric.
2. isalpha(): It produces True if characters in the given string are alphabetic.
3. isdigit(): It produces True if the characters in the given string contain only digits.
4. islower(): It produces True if the entire characters in the given string are in lowercase.
5. isupper(): It produces True if the entire characters in the given string are in uppercase.

6. istitle(): It produces True if the string is in the title case.
7. isspace(): It produces True if the string possess white-space characters only.

Example:

```
print('Vijay1990'.isalnum())
print('swati1989'.isalpha())
print('shashwat'.isalpha())
print('vimal'.isdigit())
print('56789'.isdigit())
print('abc'.islower())
print('Abc'.islower())
print('abc123'.islower())
print('ABC'.isupper())
print('Sachin is a god of cricket'.istitle())
print('Best Programming Is Python'.istitle())
print('     '.isspace())
```

```
True
False
True
False
True
True
False
True
True
False
True
True
```

Example:

```
s=input("Enter character:")
if s.isalnum():
    print("Alpha Numeric")
    if s.isalpha():
        print("Alphabet")
        if s.islower():
            print("Lower case")
        else:
            print("Upper case")
    else:
        print("Digit")
elif s.isspace():
    print("Space character")
else:       print("Non Space Special Character")
```

```
Enter character:9
Alpha Numeric
Digit
```

```
Enter character:a
Alpha Numeric
Alphabet
Lower case
```

5.8 Formatting of the Strings

Variable values are preferred for formatting the string by using the replacement operator {} and the format() method.

5.8.1 Case 1:Formatting (Default, Positional, and Keyword Arguments)

Example:

```
name   = 'vijay'
empid = 20201
exp = 4
print("{} empid {} exp {}".format(name,empid,exp))
print("{0} empid {1} exp {2}".format(name,empid,exp))
print("{x} empid {y} exp {z}".format(z=exp,y=empid,x=name))
```

```
vijay empid 20201 exp 4
vijay empid 20201 exp 4
vijay empid 20201 exp 4
```

5.8.2 Case 2: Formatting of Numbers

  f: Fixed point number(float); default precision is 6
  d: Decimal integer
  b: Binary format
  o: Octal format
  X: Hexa-decimal format in upper case
  x: Hexa-decimal format in lower case

Example:

```
print("Output of integer number : {}".format(133))
print("Output of integer number : {:d}".format(133))
print("Output of integer number : {:5d}".format(133))
print("Output of integer number : {:05d}".format(133))

Output of integer number : 133
Output of integer number : 133
Output of integer number :   133
Output of integer number : 00133
```

Example: {:07.2f}

It means there should be a minimum of seven positions, and exactly two digits are permitted after the decimal point. If it is less, then 0s will be added at the end position. If the total number is less than seven positions then 0 will be added in the most significant bit. If the total number of positions is greater than seven positions then all the integral digits will be taken care of and the only extra digit you can take is 0 (Qamar and Raza 2020).

Example:

```
print("Output of float number : {}".format(133.4565))
print("Output of float number : {:f}".format(133.3456))
print("Output of float number : {:6.2f}".format(133.5667))
print("Output of float number : {:06.3f}".format(133.5657))
print("Output of float number : {:07.4f}".format(133.56))
print("Output of float number : {:08.4f}".format(123456789.56))

Output of float number : 133.4565
Output of float number : 133.345600
Output of float number : 133.57
Output of float number : 133.566
Output of float number : 133.5600
Output of float number : 123456789.5600
```

Program: Write a program to display the decimal value in binary, octal, and hexadecimal form.

```
print("Binary:{0:b}".format(135))
print("Octal:{0:o}".format(135))
print("Hexa decimal:{0:x}".format(167))
print("Hexa decimal:{0:X}".format(167))

Binary:10000111
Octal:207
Hexa decimal:a7
Hexa decimal:A7
```

5.8.3 Case 3: Formatting for Signed Numbers

While showing the positive numbers, if you wants to include + then, write {:+d} and {:+f} and for negative numbers the– sign will be used by default (Rai 2019).

Example:

```
print("int with sign:{:+d}".format(133))
print("int with sign:{:+d}".format(-133))
print("float with sign:{:+f}".format(133.677))
print("float with sign:{:+f}".format(-133.677))

int with sign:+133
int with sign:-133
float with sign:+133.677000
float with sign:-133.677000
```

5.8.4 Case 4: Number Formatting with Alignment

The symbols used in alignment are shown in Table 5.1.

TABLE 5.1

Alignment symbols

Symbol	Purpose
<	Left alignment at the remaining space
>	Right alignment at the remaining space
^	Centre alignment at the remaining space
=	Forces the signed (+) (–) to the left most position

The right alignment is the default alignment.

Example:

```
print("{:5d}".format(18))
print("{:<5d}".format(18))
print("{:<05d}".format(18))
print("{:>5d}".format(18))
print("{:>05d}".format(18))
print("{:^5d}".format(18))
print("{:=5d}".format(-18))
print("{:^10.3f}".format(18.1234))
print("{:=8.3f}".format(-18.1234))
```

```
   18
18
18000
   18
00018
 18
-  18
  18.123
-  18.123
```

5.8.5 Case 5: String Formatting with Format()

As with numbers, you can format the string values also using format() method.

Syntax: s.format (string)

Example:

```
print("{:4d}".format(18))
print("{:4}".format("man"))
print("{:>6}".format("man"))
print("{:<4}".format("man"))
print("{:^5}".format("man"))
print("{:*^6}".format("man"))
```

```
  18
man
   man
man
 man
*man**
```

5.8.6 Case 6: Truncating Strings Using Format() Method

Example:

```
print("{:.7}".format("vijaymnnit"))
print("{:6.4}".format("vijaymnnit"))
print("{:>6.2}".format("vijaymnnit"))
print("{:^7.3}".format("vijaymnnit"))
print("{:*^7.4}".format("vijaymnnit"))
```

```
vijaymn
vija
    vi
  vij
*vija**
```

5.8.7 Case 7: Formatting Dictionary Members Using Format() Method

Example:

```
p={'id':20201,'name':'swati'}
print("Name: {x[name]} Id: {x[id]}".format(x=p))
```

```
Name: swati Id: 20201
```

Example: (to use **b)

```
p={'id':20201,'name':'swati'}
print("Name: {name}'ID: {id}".format(**p))
```

```
Name: swati'ID: 20201
```

5.8.8 Case 8: Formatting Class Members Using Format() Method

Example:

```
class Human:
    ids=20201
    name="swati"
print("Name: {x.name} Id :{x.ids}".format(x=Human()))
```

```
Name: swati Id :20201
```

Example:

```
class Human:
    def __init__(self,name,ids):
        self.name=name
        self.ids=ids
print("Name: {x.name} Id :{x.ids}".format(x=Human('vimal',20202)))
print("Name: {x.name} Id :{x.ids}".format(x=Human('Parth',20203)))
```

```
Name: vimal Id :20202
Name: Parth Id :20203
```

5.8.9 Case 9: Dynamic Formatting Using Format()

Example:

```
str1="{:{fill}{align}{width}}"
print(str1.format('man',fill='*',align='^',width=5))
print(str1.format('man',fill='*',align='^',width=7))
print(str1.format('man',fill='*',align='<',width=8))
print(str1.format('man',fill='*',align='>',width=9))
```

```
*man*
**man**
man*****
******man
```

5.8.10 Case 10: Dynamic Float Format Template

Example:

```
number="{:{align}{width}.{precision}f}"
print(number.format(133.567,align='<',width=7,precision=2))
print(number.format(133.567,align='>',width=9,precision=3))
```

```
133.57
  133.567
```

5.8.11 Case 11: Formatting Date Values

Example:

```
import datetime
d=datetime.datetime.now()
print("Date and Time Format:{:%d/%m/%Y   %H:%M:%S}".format(d))

Date and Time Format:16/01/2021   12:56:32
```

5.8.12 Case 12: Formatting Complex Numbers

Example:

```
complexNumber=1+2j
print("Real:{0.real} Imaginary :{0.imag}".format(complexNumber))

Real:1.0 Imaginary :2.0
```

Some More Examples

Program: Show how to reverse the string.

```
n = input("Enter String:")
print(n[::-1])

Enter String:vijay
yajiv
```

OR

```
n = input("Enter String:")
print(''.join(reversed(n)))

Enter String:vijay
yajiv
```

OR

```
n = input("Enter String:")
i=len(n)-1
var=''
while i>=0:
    var=var+n[i]
    i=i-1
print(var)

Enter String:vijy
yjiv
```

Program: Show how to reverse the order of words

```
n=input("Enter String:")
l=n.split()
l1=[]
i=len(l)-1
while i>=0:
    l1.append(l[i])
    i=i-1
opt=' '.join(l1)
print(opt)
```

```
Enter String:Dr. Kalam is a President of India
India of President a is Kalam Dr.
```

Program: Show how to reverse the internal content of each word.

```
s=input("Enter String:")
l=s.split()
l1=[]
i=0
while i<len(l):
    l1.append(l[i][::-1])
    i=i+1
opt=' '.join(l1)
print(opt)
```

```
Enter String:Gandhi was a honest man
ihdnaG saw a tsenoh nam
```

Program: Display the characters at odd and even position for the string.

```
n = input("Enter String:")
print("Even Position:",n[0::2])
print("Odd Position:",n[1::2])
```

```
Enter String:Ajad Hind
Even Position: Aa id
Odd Position: jdHn
```

OR

```
s=input("Enter String:")
i=0
print("Even Position:")
while i< len(s):
    print(s[i],end=',')
    i=i+2
print()
print("Odd Position:")
i=1
while i< len(s):
    print(s[i],end=',')
    i=i+2
```

```
Enter String:Ajad Hind
Even Position:
A,a, ,i,d,
Odd Position:
j,d,H,n,
```

Program: Write some Python code to merge the characters of two strings into a single string by picking the alternate characters.

```
n1=input("Enter String-1:")
n2=input("Enter String-2:")
opt=''
i,j=0,0
while i<len(n1) or j<len(n2):
    if i<len(n1):
        opt=opt+n1[i]
        i+=1
    if j<len(n2):
        opt=opt+n2[j]
        j+=1
print(opt)
```

```
Enter String-1:vimal
Enter String-2:shashwat
vsihmaaslhwat
```

Program: Write some code to sort the characters of a string and first letter symbols followed by numeric values.

```
n=input("Enter String:")
n1=n2=opt=''
for x in n:
    if x.isalpha():
        n1=n1+x
    else:
        n2=n2+x
for x in sorted(n1):
    opt=opt+x
for x in sorted(n2):
    opt=opt+x
print(opt)
```

```
Enter String:UY64TT3R88YY21
RTTUYYY1234688
```

Program: Write a program for the following:

Input: g4u2g9k5i3l2o8

Output: gggguugggggggggkkkkkiiilloooooooo

```
n=input("Enter String:")
opt=''
for x in n:
    if x.isalpha():
        opt=opt+x
        prev=x
    else:
        opt=opt+prev*(int(x)-1)
print(opt)
```

```
Enter String:g4u2g9k5i3l2o8
gggguugggggggggkkkkkiiilloooooooo
```

Program: Write some Python code for the following:

Input: d3n2p4h6

Output: dgnppthn

```python
n=input("Enter String:")
opt=''
for x in n:
    if x.isalpha():
        opt=opt+x
        prev=x
    else:
        opt=opt+chr(ord(prev)+int(x))
print(opt)
```

```
Enter String:d3n2p4h6
dgnppthn
```

Program: Write some Python code to remove the duplicate characters from the given input string.

Input: PWRRTPPYYWWGGHHHJJ

Output: PWRTYGHJ

```python
n = input("Enter String:")
l=[]
for x in n:
    if x not in l:
        l.append(x)
opt=''.join(l)
print(opt)
```

```
Enter String:PWRRTPPYYWWGGHHHJJ
PWRTYGHJ
```

Program: Write some Python code to find the number of occurrences of each character available in the given string.

Input: FASSTTFF

Output: F-3, A-1, S-2, T-2

```python
n=input("Enter String:")
d={}
for x in n:
    if x in d.keys():
        d[x]=d[x]+1
    else:
        d[x]=1
for i,j in d.items():
    print("{} = {} Times".format(i,j))
```

```
Enter String:FASSTTFF
F = 3 Times
A = 1 Times
S = 2 Times
T = 2 Times
```

Program: Write a program for the following:

Input: Deepak is Father of DM

Output: Deepak si Father fo DM

```python
s = input('Enter String:')
l = s.split()
l1 = []
i = 0
while i<len(l):
    if i%2==0:
        l1.append(l[i])
    else:
        l1.append(l[i][::-1])
    i=i+1
opt=' '.join(l1)
print('Actual String:',s)
print('Generated String:',opt)
```

```
Enter String:Deepak is Father of DM
Actual String: Deepak is Father of DM
Generated String: Deepak si Father fo DM
```

5.9 Conclusion

In this chapter, you have learned about working with strings, which are objects that contain sequences of character data. Processing the character data is an integral part of programming. After studying this chapter, the reader will be able to understand the full in-depth analysis of defining, performing logical operations, and defining every part of a string for useful analysis.

Review Questions

1. How do you create a string using a constructor of the str class?
2. Write a procedure to traverse every fourth character of a string.
3. How is a subset of string obtained?
4. How can a string be broken?
5. Explain the index operator.

Programming Assignments

PA 1: Write a program that takes an input string containing binary digits and converts it into an equivalent decimal integer.

PA 2: Write a function count_letter(word, letter) that takes a word and a letter as arguments and returns the number of occurrences of that letter in the word.

PA 3: Write a program that traverses all the elements using a while loop.

PA 4: Write a program that converts a hexadecimal number into its equivalent binary number.

References

Campbell, Matthew. 2014. *Objective-C quick syntax reference*. Apress.

Celko, Joe. Joe. 2010. *Celko's SQL for smarties: advanced SQL programming*. Elsevier.

Chivers, Ian D., and Jane Sleightholme. 2018. *Introduction to programming with Fortran*. Vol. 2. Springer.

Huff, Brian. 2006. *The definitive guide to Stellent content server development*. Apress.

Qamar, Usman, and Muhammad Summair Raza. 2020. Data science programming languages. In *Data science concepts and techniques with applications*, pp. 153–196. Springer, Singapore.

Rai, Laxmisha, ed. 2019. *Programming in C++: Object oriented features*. Vol. 5. Walter de Gruyter GmbH & Co KG.

Scholtz, Jean, and Susan Wiedenbeck. 1990. Learning second and subsequent programming languages: A problem of transfer. *International Journal of Human-Computer Interaction* 2, no. 1: 51–72.

Stine, James E. 2004. *Digital computer arithmetic data path design using verilog HDL*. Springer Science & Business Media.

Walters, Gregory. 2014. *Strings In The Python Quick Syntax Reference*. Apress: 25–42.

Yalagi, Pratibha S., Trupti S. Indi, and Manisha A. Nirgude. 2016. Enhancing the cognitive level of novice learners using effective program writing skills. International Conference on Learning and Teaching in Computing and Engineering (LaTICE). IEEE: 167–171.

6

Data Structures in Python

6.1 Introduction

Data structure deals with how the data is organized and held in the memory when the program processes it. Python is used worldwide in various fields like deploying websites, artificial intelligence, machine learning, and many more. For making all such tasks possible, data plays a vital role, which means that data should be stored efficiently and accessed in a timely manner. For achieving this, data structures are the only key. Python has implicit support for data structures, which enables you to store and access data. These data structures are called list, tuple, dictionary, and set. Python also allows its users to create their own data structures enabling them to have full control over their functionality.

6.2 List

A user may need to store variables of the same data type at different points. Using a list, a programmer can use a single variable to store all the elements of the same or different data type and even print them. In Python, a list is a sequence of values called items or elements. The elements can be of any type. A list is mutable in nature; the contents can be changed. The insertion order remains preserved while performing operations on a list. It permits the heterogeneous and duplicate objects in it (Kachroo and Özbay 1999). On the basis of our requirement, the user can increase or decrease its size; thus it is dynamic in nature. In a list, the elements are placed within a square bracket, differentiated by a comma separator.

Example: [10, "P", "S", 10, 20, 30]

−6	−5	−4	−3	−2	−1
10	P	S	10	20	30
0	1	2	3	4	5

6.2.1 Creation of List Objects

The list class defines lists. A programmer can use a list's constructor to create a list. The empty list object can be created as follows:

Example:

```
list=[]
print(list)
print(type(list))

[]
<class 'list'>
```

6.2.1.1 With Dynamic Input

Example:

```
list=eval(input("Enter List:"))
print(list)
print(type(list))

Enter List:[30,50,60,70]
[30, 50, 60, 70]
<class 'list'>
```

6.2.1.2 With List() Function

Example:

```
l = list(range(0,8,2))
print(l)
print(type(l))

[0, 2, 4, 6]
<class 'list'>
```

Example:

```
x="vijay"
l=list(x)
print(l)

['v', 'i', 'j', 'a', 'y']
```

6.2.1.3 With Split() Function

Example:

```
x="Teaching is best job"
l=x.split()
print(l)
print(type(l))

['Teaching', 'is', 'best', 'job']
<class 'list'>
```

6.2.2 Accessing Elements of a List

The elements of a list are unidentifiable by their positions. Hence, the index[] operator is used to access them.

6.2.2.1 By Using an Index

The elements of a list start with a zero index. A list follows a positive as well as negative index. A positive index is for a left-to-right list and a negative index is for right-to-left (Walters 2014).

Example:

```
1 =[11,22,33,44]
print(1[0])
print(1[-1])
print(1[10])

11
44

IndexError: list index out of range
```

6.2.2.2 By Using a Slice Operator

In slicing, the first two parameters are the start index and end index. Thus, we need to add a third parameter as the step size to select a list with step size.
Syntax: listb = lista[start_index: stop_index: step_value]

Example:

```
n=[10,20,30,40,50,60,70,80,90,100]
print(n[2:6:3])
print(n[5::2])
print(n[3:7])
print(n[9:2:-2])
print(n[5:100])

[30, 60]
[60, 80, 100]
[40, 50, 60, 70]
[100, 80, 60, 40]
[60, 70, 80, 90, 100]
```

6.2.3 List vs Mutability

After the creation of a list, you can modify its contents; hence it is mutable in nature. With mutability any changes or modifications can be performed as and when required by the user so after declaring a list its contents can still be modified.

Example:

```
1=[100,200,300,400]
print(1)
1[1]=500
print(1)

[100, 200, 300, 400]
[100, 500, 300, 400]
```

6.2.4 Traversing the Elements of a List

Traversal means to access the elements of a list sequentially.

6.2.4.1 *By Using the While Loop*

Example:

```
l = [10,20,30,40,50]
i = 0
while i < len(l):
    print(l[i])
    i=i+1
```

```
10
20
30
40
50
```

6.2.4.2 *By Using the For Loop*

Example:

```
l = [10,20,30,40,50]
i = 0
for i in l:
    print(i)
```

```
10
20
30
40
50
```

Program: Write some Python code to display even numbers from the entered list.

```
l=[11,20,39,40,50]
for n1 in l:
    if n1%2==0:
        print(n1)
```

```
20
40
50
```

6.2.5 Important Functions of a List

6.2.5.1 *To Get Information about a List*

a) len(): This function is used to return the quantity of elements present in the list.

Example:

```
p = [11,22,33,44,55]
print(len(p))
```

```
5
```

Program: Write a program to display the elements index wise using the len() function.

```
l = ["VIZ","QIZ","RIZ"]
n = len(l)
for i in range(n):
    print(l[i],"Positive index: ",i,"Negative index: ",i-n)
```

```
VIZ Positive index:  0 Negative index:  -3
QIZ Positive index:  1 Negative index:  -2
RIZ Positive index:  2 Negative index:  -1
```

b) count(): This function is used to return the count of specified numbers in the list.

Example:

```
l=[10,40,66,66,66,40]
print(l.count(10))
print(l.count(40))
print(l.count(66))
```

```
1
2
3
```

c) index(): This function is used to return the index of the first occurrence of the mentioned item.

Example:

```
l = [10, 30, 30, 30, 20, 30, 20]
print(l.index(10))
print(l.index(20))
print(l.index(30))
```

```
0
4
1
```

6.2.5.2 Manipulating Elements of a List

a) append() Function: This function is used to add new items at the end of a list.

Example:

```
l=[]
l.append("P")
l.append("Q")
l.append("R")
print(l)
```

```
['P', 'Q', 'R']
```

Program: Write some Python code to sum all the elements to list up to 30, which are divisible by 5.

```
l=[]
for i in range(30):
    if i%5==0:
        l.append(i)
print(l)
```

```
[0, 5, 10, 15, 20, 25]
```

b) insert() Function: It is used to insert an item at a specified index position.

Example:

```
l=[10,20,30,40,50]
l.insert(2,70)
print(l)
```

```
[10, 20, 70, 30, 40, 50]
```

Example:

```
l=[10,20,30,40,50]
l.insert(10,70)
l.insert(-10,90)
print(l)

[90, 10, 20, 30, 40, 50, 70]
```

The basic difference between append() and insert() is shown in Table 6.1.

TABLE 6.1
Difference between append() and insert()

append()	insert()
This function by default adds any element at the end of the list.	This function adds any element at the specified index number of the list.

c) extend() Function: If you wish to add the items of a list into another list, then the extend() function is preferred.

Example:

```
r1=["X","Y","Z"]
r2=["A","B","C"]
r1.extend(r2)
print(r1)

['X', 'Y', 'Z', 'A', 'B', 'C']
```

Example:

```
r1=["X","Y","Z"]
r1.extend("MOON")
print(r1)

['X', 'Y', 'Z', 'M', 'O', 'O', 'N']
```

d) remove() Function: If you wish to remove particular items from the list then the remove() function is preferred.

Example:

```
l=[80,100,60,80,90]
l.remove(80)
print(l)

[100, 60, 80, 90]
```

If the mentioned item is not available in the list then an error message will be displayed.

Example:

```
l=[80,100,60,80,90]
l.remove(10)
print(l)

ValueError: list.remove(x): x not in list
```

e) pop() Function: If you wish to remove or return the last element of the list then the pop() function is preferred.

Example:

```
1=[100,300,30,50,60]
print(1.pop())
print(1.pop())
print(1)

60
50
[100, 300, 30]
```

The pop() function will raise an error message if the list is empty.

Example:

```
1 = []
print(1.pop())

IndexError: pop from empty list
```

If you wish to remove or return the specified element from the list then, declare as follows: n.pop(index)

Example:

```
1 = [10,20,30,40,50]
print(1.pop())
print(1.pop(1))
print(1.pop(10))

50
20

IndexError: pop index out of range
```

The difference between the remove() function and pop() function is shown in Table 6.2.

TABLE 6.2
Difference between remove() and pop()

remove()	pop()
This function is used to remove a special element from the list.	This function is used to remove the last element from the list.
It doesn't return any value.	It returns a removed element.

6.2.5.3 Ordering Elements of a List

a) reverse(): It reverses the elements of the list.

Example:

```
1=[20,40,60,80]
1.reverse()
print(1)

[80, 60, 40, 20]
```

b) sort(): It sorts the elements of the list. To use the sort() function, the list must comprise homogeneous elements only or an error message will be raised. By default, the sorting order for numbers is always the ascending order whereas for strings the by-default sorting order is always the alphabetical order.

Example:

```
l = [30,6,14,18,1]
l.sort()
print(l)
l1 = ["Man","Van","Lan","App"]
l1.sort()
print(l1)

[1, 6, 14, 18, 30]
['App', 'Lan', 'Man', 'Van']
```

Example:

```
l=[20,10,"A","B"]
l.sort()
print(l)

TypeError: '<' not supported between instances of 'str' and 'int'
```

Example:

```
l=[30,"R",70,"D"]
l.sort()
print(l)

TypeError: '<' not supported between instances of 'str' and 'int'
```

The list can be sorted in reverse order using (reverse=True) argument.

Example:

```
l = [4,1,3,2,7,0]
l.sort()
print(l)
l.sort(reverse = True)
print(l)
l.sort(reverse = False)
print(l)

[0, 1, 2, 3, 4, 7]
[7, 4, 3, 2, 1, 0]
[0, 1, 2, 3, 4, 7]
```

6.2.6 Aliasing and Cloning of List Objects

Aliasing is the technique of giving another reference variable to the already existing list. The slice operator and copy() function are used for cloning a list.

Example:

```
x=[30,250,300,460]
y=x
print(id(x))
print(id(y))

3000475114048
3000475114048
```

The issue of using one reference variable is that if any changes are performed in its content then those changes will be reflected in another reference variable.

Example:

```
x = [1,23,30,46]
y = x
y[1] = 80
print(x)
```

```
[1, 80, 30, 46]
```

6.2.6.1 By Using the Slice Operator

Example:

```
a = [100,200,300,400]
b = a[:]
b[2] = 90
print(a)
print(b)
```

```
[100, 200, 300, 400]
[100, 200, 90, 400]
```

6.2.6.2 By Using the Copy() Function

Example:

```
a = [100,200,300,400]
b = a.copy()
b[2] = 90
print(a)
print(b)
```

```
[100, 200, 300, 400]
[100, 200, 90, 400]
```

6.2.7 Using Mathematical Operators for List Objects

Two operators are used for list objects; + and *

6.2.7.1 Concatenation Operator(+)

The concatenation operator is used to concatenate two lists. For using the + operator, both the arguments should be the list objects or an error message will be raised.

Program: Write a program to concatenate two different lists.

```
a = [100, 200, 300]
b = [400, 500, 600]
c = a+b
print(c)
```

```
[100, 200, 300, 400, 500, 600]
```

6.2.7.2 Repetition Operator(*)

The multiplication operator is used to replicate the elements of the list.

Example:

```
a = [10, 20, 30]
b = a*3
print(b)
```

```
[10, 20, 30, 10, 20, 30, 10, 20, 30]
```

6.2.8 Comparing List Objects

The comparison operators are used for list objects. Whenever you are using the relational operators among the list objects, the first element comparison will be done (Srinivasa and Srinidhi 2018).

Program: Write a program to compare the equality of three different lists.

```
a = ["Man", "Wan", "Lan"]
b = ["Man", "Wan", "Lan"]
c = ["MAN", "WAN", "LAN"]
print(a == b)
print(a == c)
print(a != c)
```

```
True
False
True
```

Example:

```
p = [50, 20, 30]
q = [40, 50, 60, 100, 200]
print(p>q)
print(p>=q)
print(p<q)
print(p<=q)
```

```
True
True
False
False
```

Example:

```
p = ["Man", "Wan", "Lan"]
q = ["Man", "Wan", "Lan"]
print(p>q)
print(p>=q)
print(p<q)
print(p<=q)
```

```
False
True
False
True
```

6.2.9 Membership Operators

Membership operators are used to check whether the element is a member of the list or not. The allowed membership operators are:

- in operator
- not in operator

Example:

```
l=[10,20,30,40]
print (30 in l)
print (10 not in l)
print (60 in l)
print (80 not in l)

True
False
False
True
```

6.2.10 Clear() Function

The clear() function is used to remove all the elements from the list.

Example:

```
l=[100,200,300,400]
print(l)
l.clear()
print(l)

[100, 200, 300, 400]
[]
```

6.2.11 Nested List

A nested list contains one list inside another list.

Example:

```
l=[1,2,[3,4]]
print(l)
print(l[0])
print(l[2])
print(l[2][0])
print(l[2][1])

[1, 2, [3, 4]]
1
[3, 4]
3
4
```

Program: Write a program to print the data row wise and in matrix style of a given list.

```
l=[[10,20,30],[40,50,60],[70,80,90]]
print(l)
print("Row Data:")
for r in l:
    print(r)
print("Matrix:")
for i in range(len(l)):
    for j in range(len(l[i])):
        print(l[i][j],end=' ')
    print()

[[10, 20, 30], [40, 50, 60], [70, 80, 90]]
Row Data:
[10, 20, 30]
[40, 50, 60]
[70, 80, 90]
Matrix:
10 20 30
40 50 60
70 80 90
```

6.2.12 List Comprehensions

List comprehension is used to create a new list from existing sequences. It is a tool for transforming a given list into another list.

Syntax: list = [expression in list if condition]

Example:

```
p = [ x*x for x in range(1,11)]
print(p)
q = [2**x  for x in range(1,6)]
print(q)
r = [x for x in p if x%2==0]
print(r)

[1, 4, 9, 16, 25, 36, 49, 64, 81, 100]
[2, 4, 8, 16, 32]
[4, 16, 36, 64, 100]
```

Example:

```
s=["vijay","swati","vimal","parth"]
l=[w[0] for  w in s]
print(l)

['v', 's', 'v', 'p']
```

Example:

```
l1=[10,20,30,40]
l2=[30,40,50,60]
l3=[i for i in l1 if i not in l2]
print(l3)
l4=[i for i in l1 if i in l2]
print(l4)

[10, 20]
[30, 40]
```

Example:

```
s="Dravid is The wall".split()
print(s)
l=[[w.upper(),len(w)]  for w in s]
print(l)

['Dravid', 'is', 'The', 'wall']
[['DRAVID', 6], ['IS', 2], ['THE', 3], ['WALL', 4]]
```

Program: Write some Python code to print the different vowels available in the given word.

```
v=['a','e','i','o','u']
w=input("Enter the String:")
f=[]
for l in w:
    if l in v:
        if l not in f:
            f.append(l)
print(f)
print("Vowels",w,"is",len(f))

Enter the String:swatiresearchscholor
['a', 'i', 'e', 'o']
Vowels swatiresearchscholor is 4
```

6.3 Tuple

A tuple contains a sequence of items of many types. The elements of tuples are fixed. Once a tuple has been created, we cannot add or delete elements, or even shuffle their order. Hence, tuples are immutable. This means that once created, they cannot be changed. Since tuples are immutable, their length is also fixed (McKinney 2010). A new tuple must be created to grow or shrink an earlier one.

Example:

```
t=30,60,80,90
print(t)
print(type(t))
t1=()
print(type(t1))

(30, 60, 80, 90)
<class 'tuple'>
<class 'tuple'>
```

In case of a special-valued tuple, the value must terminate with a comma or it will not be regarded as a tuple.

Example:

```
t=(40)
print(t)
print(type(t))

40
<class 'int'>
```

Example:

```
t=(40,)
print(t)
print(type(t))

(40,)
<class 'tuple'>
```

6.3.1 Tuple Creation

A tuple is an inbuilt data type in Python. In order to create a tuple, the elements of tuples are enclosed in parenthesis instead of square brackets.

1. p = ():   An empty tuple will be created.
2. p = (22,)

   p = 22,   A single-valued tuple will be created. Here, parentheses are optional.

3. p = 11, 22, 33

   p = (11, 22, 33)   A multi-valued tuple will be created. Here, parentheses are optional.

4. By using the tuple() function:

Example:

```
l=[100,200,300]
t=tuple(l)
print(t)
t=tuple(range(10,18,2))
print(t)
```

```
(100, 200, 300)
(10, 12, 14, 16)
```

6.3.2 Accessing Elements of a Tuple

The elements of a tuple can be accessed by the index and slice operator.

6.3.2.1 By Using the Index

Example:

```
t = (10,20,30,40,50)
print(t[0])
print(t[-1])
print(t[100])
```

```
10
50
```

```
IndexError: tuple index out of range
```

6.3.2.2 By Using the Slice Operator

Example:

```
t=(10,20,30,40,50)
print(t[2:6])
print(t[3:50])
print(t[::4])
```

```
(30, 40, 50)
(40, 50)
(10, 50)
```

6.3.3 Tuple vs Immutability

A tuple is immutable in nature; you cannot modify the contents of a tuple once created.

Example:

```
t = (12,20,40,60)
t[1] = 70
print(t)
```

```
TypeError: 'tuple' object does not support item assignment
```

6.3.4 Mathematical Operators for a Tuple

1. Concatenation Operator (+):

Example:

```
t1=(10,20,30)
t2=(40,50,60)
t3=t1+t2
print(t3)
```

```
(10, 20, 30, 40, 50, 60)
```

2. Multiplication or Repetition Operator (*)

Example:

```
t1=(100,200)
t2=t1*3
print(t2)
```

```
(100, 200, 100, 200, 100, 200)
```

6.3.5 Important Functions of a Tuple

a) len(): It returns the number of elements in a tuple.

Example:

```
t = (10,20,30,40)
print(len(t))
```

```
4
```

b) count(): It returns the quantity of occurrences of an element t.

Example:

```
t = (30, 20, 30, 30, 20)
print(t.count(30))
```

```
3
```

c) index(): It returns the index of element t.

Example:

```
t = (100, 200, 100, 100, 200)
print(t.index(100))
print(t.index(200))
```

```
0
1
```

d) sorted(): It is preferred for sorting the elements on the basis of the default sorting order.

Example:

```
t=(400,100,300,200)
t1=sorted(t)
print(t1)
print(t)
```

```
[100, 200, 300, 400]
(400, 100, 300, 200)
```

In a tuple, the data can be sorted in decreasing order using (reverse = True), as shown below:

Example:

```
t1 = sorted(t, reverse = True)
print(t1)

[400, 300, 200, 100]
```

e) min() And max() Functions: It returns the element with the smallest and the greatest value.

Example:

```
t = (400,100,300,200,38)
print(min(t))
print(max(t))

38
400
```

f) cmp(): It compares the elements of both the tuples (Goodrich, Tamassia, and Goldwasser 2013). This function is present only in Python 2; Python 3 doesn't support this function.
  - If the tuples are the same then it will return 0.
  - If tuple1 is less than tuple2 then it will return −1.
  - If tuple1 is greater than tuple2 then it will return +1.

Example:

```
t1=(10,20,30)
t2=(40,50,60)
t3=(10,20,30)
print(cmp(t1,t2))
print(cmp(t1,t3))
print(cmp(t2,t3))

NameError: name 'cmp' is not defined
```

6.3.6 Tuple Packing and Unpacking

A tuple can be created by packing the set of variables.

Example: (w, x, y, z are packed into a tuple t)

```
x = 10
y = 20
z = 30
w = 40
t = x,y,z,w
print(t)

(10, 20, 30, 40)
```

The reverse of tuple packing is tuple unpacking. The tuple can be unpacked, and its values can be assigned to other variables.

Example:

```
t=(33,22,39,47)
w,x,y,z=t
print("w=",w,"x=",x,"y=",y,"z=",z)

w= 33 x= 22 y= 39 z= 47
```

While unpacking the tuple, the number of values and the variables must be same or an error message will be raised.

Example:

```
t = (1,2,3,4)
x,y,z = t

ValueError: too many values to unpack (expected 3)
```

6.3.7 Tuple Comprehension

Python does not support tuple comprehension. It will produce the generator object.

Example:

```
t= ( a**2  for a in range(1,5))
print(type(t))
for a in t:
    print(a)

<class 'generator'>
1
4
9
16
```

Program: Write some code to put numbers in a tuple and display their sum and average.

```
t=eval(input("Enter the values:"))
l=len(t)
sum=0
for i in t:
    sum=sum+i
print("Sum=",sum)
print("Average=",sum/l)

Enter the values:(30,40,50,60)
Sum= 180
Average= 45.0
```

Thus, the notified differences between a list and a tuple are discussed in Table 6.3.

TABLE 6.3

Difference between a list and tuple

List	Tuple
A list is a set of comma-separated values mentioned in the square brackets. Here, the brackets are mandatory.	A tuple is a set of comma-separated values mentioned in the parentheses. Here, the parentheses are optional.
A list is mutable in nature once a list is created, any number of changes can be performed on it.	A tuple is immutable in nature; once a tuple is created, you cannot change its content.
If the elements of the list are not fixed, then one should go for list.	If the elements are fixed and predefined, then a tuple is good.
The list objects cannot be used in dictionaries as keys because the keys in dictionaries are immutable and hashable.	The tuple objects cannot be used in dictionaries as keys because the keys in dictionaries are immutable and hashable.

6.4 Set

A set is an unordered collection of unique elements without duplicates. A set is mutable. Hence, we can easily add or remove elements from a set. The set data structure in Python is used to support mathematical set operations.

Example:

```
s={1,2,3,4}
print(s)
print(type(s))

{1, 2, 3, 4}
<class 'set'>
```

6.4.1 Creation of Set Objects

A programmer can create a set by enclosing the elements inside a pair of curly brackets{}. The elements within the set can be separated by commas. We can also create a set using the inbuilt set() function or from an existing list or tuple.

Example:

```
l = [1,2,5,5,3,3,4,1,1,2,1]
s=set(l)
print(s)

{1, 2, 3, 4, 5}
```

Example:

```
s=set(range(5))
print(s)

{0, 1, 2, 3, 4}
```

Example:

```
s={}
print(s)
print(type(s))

{}
<class 'dict'>
```

Example:

```
s=set()
print(s)
print(type(s))

set()
<class 'set'>
```

6.4.2 Important Functions of a Set

a) add(x): It adds the element x to an existing set.

Example:

```
s={100,300,500}
s.add(400);
print(s)

{400, 100, 500, 300}
```

b) update(x,y,z): It is used to add number of items to the set. All the items available in the iterable objects are added to the set.

Example:

```
s={100,200,300}
l=[400,500,600,100]
s.update(l,range(5))
print(s)

{0, 1, 2, 3, 100, 4, 200, 300, 400, 500, 600}
```

The differences between add() and update() functions of a set are listed in Table 6.4.

TABLE 6.4
Difference between add() and update()

add()	update()
This function is used to add individual items to the set.	This function is used to add multiple items in the set.
It takes only one argument as an iterable object	It can take multiple arguments as iterable objects.

c) copy(): It produces the coned copy of the set.

Example:

```
s = {1,2,3}
s1 = s.copy()
print(s1)
```

```
{1, 2, 3}
```

d) pop(): It deletes and returns the arbitrary element from the set.

Example:

```
s={4,1,3,2}
print(s)
print(s.pop())
print(s)
```

```
{1, 2, 3, 4}
1
{2, 3, 4}
```

e) remove(x): It deletes the specified element from the set and if that element is not available in the set, an error message will be raised.

Example:

```
s = {4, 1, 3, 2}
s.remove(3)
print(s)
s.remove(5)
print(s)
```

```
{1, 2, 4}
```

```
KeyError: 5
```

f) discard(x): It deletes the specified element from the set and if that element is not available, no error message will be raised.

Example:

```
s = {1, 2, 3,7}
s.discard(1)
print(s)
s.discard(5)
print(s)
```

```
{2, 3, 7}
{2, 3, 7}
```

g) clear(): It is used to remove all elements from the set.

Example:

```
s={1,2,3}
print(s)
s.clear()
print(s)
```

```
{1, 2, 3}
set()
```

6.4.3 Mathematical Operations on a Set

a) union(): The union of two sets A and B is a set of elements that are in A, in B or in both A and B.

 x.union(y) or x | y: It returns all the elements of set x and y.

Example:

```
p = {1, 2, 3, 4}
q = {3, 4, 5, 6}
print (p.union(q))
print (p|q)
```

```
{1, 2, 3, 4, 5, 6}
{1, 2, 3, 4, 5, 6}
```

b) intersection(): The intersection of two sets A and B is the set that contains all the elements of A that also belong to B.

 x.intersection(y) OR x&y: It gives the common elements of set x and y.

Example:

```
p = {1, 2, 3, 4}
q = {3, 4, 5, 6}
print (p.intersection(q))
print(p&q)
```

```
{3, 4}
{3, 4}
```

c) difference(): The difference between two sets A and B is the set that contains the element in set A but not in set B.

 x.difference(y) OR x-y: It gives the elements of set x that are not available in set y.

Example:

```
p = {1, 2, 3, 4}
q = {3, 4, 5, 6}
print (p.difference(q))
print (p-q)
print (q-p)
```

```
{1, 2}
{1, 2}
{5, 6}
```

d) symmetric_difference(): The symmetric difference is a set that contains elements from either set but not in both sets.

x.symmetric_difference(y) ORx^y: It gives the distinct elements of set x and y.

Example:

```
p = {1, 2, 3, 4}
q = {3, 4, 5, 6}
print (p.symmetric_difference(q))
print(p^q)
```

```
{1, 2, 5, 6}
{1, 2, 5, 6}
```

Membership operators

1. in operator
2. not in operator

Example:

```
s=set("vijay")
print(s)
print('v' in s)
print('z' in s)
```

```
{'v', 'i', 'j', 'a', 'y'}
True
False
```

6.4.4 Set Comprehension

Set comprehension is possible. The indexing and slicing operations are not supported by set objects.

Example:

```
s = {x*x  for x in range(5)}
print (s)
s = {2**x for x in range(2,10,2)}
print (s)
```

```
{0, 1, 4, 9, 16}
{16, 256, 64, 4}
```

Example:

```
s = {1,2,3,4}
print(s[0])
print(s[1:3])
```

```
TypeError: 'set' object is not subscriptable
```

Program: Write a program to remove the redundant elements from the list.

```
l = [1,2,5,5,3,3,4,1,1,2,1]
s=set(l)
print(s)
```

```
{1, 2, 3, 4, 5}
```

Program: Write a program to display distinct vowels of a given word.

```
w=input("Enter String:")
s=set(w)
v={'a','e','i','o','u'}
d=s.intersection(v)
print("Vowels",w,":",d)
```

```
Enter String:swatiresearchscholoru
Vowels swatiresearchscholoru : {'u', 'e', 'o', 'a', 'i'}
```

6.5 Dictionary

In Python, a dictionary is a collection that stores values along with keys. The sequence of such keys and value pairs is separated by commas. These pairs are sometimes called entries. If you wish to show the set of objects as a (key, value) pair, then dictionaries are preferred.

Example:

>   roll_no----name
>   phone_no----address
>   ip_address----domain_name

In a dictionary, duplicate values allowed.

6.5.1 Creating a Dictionary

We can create a dictionary by enclosing the items inside a pair of curly brackets{}. One way to start a dictionary is to create an empty dictionary first and then add items to it.

Example:

>   d = {} OR d = dict()

We can add entries as follows:

Example:

```
d={}
d[10]="vijay"
d[20]="swati"
d[30]="vimal"
print(d)
```

```
{10: 'vijay', 20: 'swati', 30: 'vimal'}
```

6.5.2 Accessing Data from a Dictionary

The data can be accessed using keys.

Example: (Display of error message when the mentioned key is not present)

```
d = {10:'vijay',20:'swati', 30:'vimal'}
print(d[10])
print(d[30])
print(d[50])
```

```
vijay
vimal

KeyError: 50
```

Program: Write a program to print the name and the percentage marks in a dictionary.

```
d={}
n1=int(input("Enter number: "))
i=1
while i <=n1:
    n=input("Enter Name: ")
    m=input("Enter Percentage: ")
    d[n]=m
    i=i+1
print("Name","\t","Percentage")
for x in d:
    print(x,"\t",d[x])
```

```
Enter number: 2
Enter Name: Shashwat
Enter Percentage: 70
Enter Name: Parth S
Enter Percentage: 80
Name        Percentage
Shashwat            70
Parth S             80
```

6.5.3 Updating a Dictionary

If you wish to add any new (key, value) pair in the already defined dictionary, then the following syntax is used:

Syntax: d[key]=value

On the contrary, if the key is already present then the value will be updated.

Example:

```
d={10:"vijay",20:"vimal",30:"parth"}
print(d)
d[40]="swati"
print(d)
d[10]="shashwat"
print(d)

{10: 'vijay', 20: 'vimal', 30: 'parth'}
{10: 'vijay', 20: 'vimal', 30: 'parth', 40: 'swati'}
{10: 'shashwat', 20: 'vimal', 30: 'parth', 40: 'swati'}
```

6.5.4 Deleting Elements from a Dictionary

a) del→d[key]

It will delete the entry of the corresponding key but if the key is not present then an error message will be raised.

Example:

```
d={10:"vijay",20:"vimal",30:"swati"}
print(d)
del d[10]
print(d)
del d[40]

{10: 'vijay', 20: 'vimal', 30: 'swati'}
{20: 'vimal', 30: 'swati'}

KeyError: 40
```

b) d.clear(): It is used to remove all entries from the dictionary.

Example:

```
d={10:"vijay",20:"vimal",30:"swati"}
print(d)
d.clear()
print(d)

{10: 'vijay', 20: 'vimal', 30: 'swati'}
{}
```

c) del d: It will delete the dictionary object then you cannot access that object.

Example:

```
d={10:"vijay",20:"vimal",30:"swati"}
print(d)
del d
print(d)

{10: 'vijay', 20: 'vimal', 30: 'swati'}

NameError: name 'd' is not defined
```

6.5.5 Important Functions of a Dictionary

a) len(): It gives the quantity of elements present in the dictionary.
b) clear(): It will delete the complete elements from dictionary.
c) get(): It will return the associated value of the key if it is present or none will be returned.

Syntax: d.get(key)

If the key is present then it will return the associated value, otherwise the default value will be returned.

Syntax: d.get(key, defaultvalue)

Example:

```
d={10:"vijay",20:"vimal",30:"swati"}
print(d[10])
print(d.get(10))
print(d.get(40))
print(d.get(10,"Shashwat"))
print(d.get(40,"Parth"))
print(d[40])

vijay
vijay
None
vijay
Parth

KeyError: 40
```

e) pop():It deletes the value of the specified key.

Syntax: d. pop(key)

Example:

```
d={10:"vijay",20:"vimal",30:"swati"}
print(d.pop(10))
print(d)
print(d.pop(40))

vijay
{20: 'vimal', 30: 'swati'}

KeyError: 40
```

f) popitem(): It removes a random item (key-value) from the dictionary.

Example:

```
d={10:"vijay",20:"vimal",30:"swati"}
print(d)
print(d.popitem())
print(d)

{10: 'vijay', 20: 'vimal', 30: 'swati'}
(30, 'swati')
{10: 'vijay', 20: 'vimal'}
```

While deleting an element from the empty dictionary, an error message will be raised.

Example:

```
d= {}
print(d.popitem())

KeyError: 'popitem(): dictionary is empty'
```

g) keys(): It returns all keys associated with a dictionary.

Example:

```
d={10:"vijay",20:"vimal",30:"swati"}
print(d.keys())
for i in d.keys():
    print(i)

dict_keys([10, 20, 30])
10
20
30
```

h) values(): All the values of the dictionary will be returned.

Example:

```
d={10:"vijay",20:"vimal",30:"swati"}
print(d.values())
for i in d.values():
    print(i)

dict_values(['vijay', 'vimal', 'swati'])
vijay
vimal
swati
```

i) items():It will return the list of tuples showing the (key-value) pairs.

$$\left[(k,v),(k,v),(k,v)\right]$$

Example:

```
d={10:"vijay",20:"vimal",30:"swati"}
for i,j in d.items():
    print(i,"==>",j)

10 ==> vijay
20 ==> vimal
30 ==> swati
```

j) copy(): It is used to construct the cloned copy. If the key is present then it will produce the associated value whereas if the key is not present then the mentioned (key-value) is appended as a new element to the dictionary (Miller and Ranum 2011).

```
p = d.copy();
    setdefault():
    d.setdefault(key,value)
```

Example:

```
d={10:"vijay",20:"vimal",30:"swati"}
print(d.setdefault(40,"shashwat"))
print(d)
print(d.setdefault(10,"Parth"))
print(d)
```

```
shashwat
{10: 'vijay', 20: 'vimal', 30: 'swati', 40: 'shashwat'}
vijay
{10: 'vijay', 20: 'vimal', 30: 'swati', 40: 'shashwat'}
```

k) update(): The elements available in the dictionary x will be appended to the given dictionary d.

Syntax: d.update(x)

Program: Write a program to take a dictionary and display the total of the taken values.

```
d=eval(input("Enter Data:"))
x=sum(d.values())
print("Addition= ",x)
```

```
Enter Data:{"vij":1,"vim":4,"sha":6}
Addition=  11
```

Program: Write a program to identify the number of occurrences of each letter in the string.

```
w=input("Enter input:")
d={}
for a in w:
    d[a]=d.get(a,0)+1
for i,j in d.items():
    print(i,"==> ",j," times")
```

```
Enter input:addition
a ==>  1   times
d ==>  2   times
i ==>  2   times
t ==>  1   times
o ==>  1   times
n ==>  1   times
```

Program: Write a program to identify the number of occurrences of each vowel available in the string.

```
w=input("Enter input: ")
v={'a','e','i','o','u'}
d={}
for a in w:
    if a in v:
        d[a]=d.get(a,0)+1
for i,j in sorted(d.items()):
    print(i,"==> ",j," times")
```

```
Enter input: percentagepoint
a ==>  1   times
e ==>  3   times
i ==>  1   times
o ==>  1   times
```

Program: Write a program to input student names and marks and create its dictionary and print student marks by inputting student names.

```
x=int(input("Enter the number: "))
d={}
for i in range(x):
    n=input("Enter Name: ")
    m=input("Enter Marks: ")
    d[n]=m
while True:
    n=input("Enter Name for find Marks: ")
    m=d.get(n,-1)
    if m== -1:
        print("Not Found Information")
    else:
        print(n,"==>",m)
    opt=input("Do you want to find another information [Yes|No]")
    if opt=="No":
        break
print("Thank you...")
```

```
Enter the number: 2
Enter Name: swati
Enter Marks: 90
Enter Name: vijay
Enter Marks: 80
Enter Name for find Marks: swati
swati ==> 90
Do you want to find another information [Yes|No]No
Thank you...
```

6.5.6 Dictionary Comprehension

The comprehension concept is applicable in dictionaries.

Example:

```
s={i:i*i for i in range(1,5)}
print("Squares",s)
d={i:2*i for i in range(1,5)}
print("Doubles",d)
```

```
Squares {1: 1, 2: 4, 3: 9, 4: 16}
Doubles {1: 2, 2: 4, 3: 6, 4: 8}
```

6.6 Conclusion

Data structures are the essential building blocks that are used to organize all the digital information. So, in this chapter readers have explored the scope of data, importance of data structures, how, where, and when to use them efficiently in your program, as the data structures are the backbone of any program.

Review Questions

1. How is a tuple created?
2. Compare a list and tuple.
3. List the in-built functions supported by a tuple.
4. How is indexing and slicing of tuples done?
5. Which operator is used to access the elements in a tuple?
6. List the inbuilt functions supported by sets.
7. What is the use of the len() function?
8. Which operator is used to access the elements in a set?
9. List the inbuilt functions supported by a dictionary.
10. Mention the properties of keys in dictionaries.
11. How are the elements of a list reversed?
12. What is the benefit of step size in a list?
13. List the inbuilt functions supported by sets.
14. Which operator is used to access the elements in a set?

Programming Assignments

PA 1: Consider a tuple T=(10, 20, 30, 40, 50, 60). Write a program to save the values present at odd indexes into a new tuple.

PA 2: Write a program to traverse tuples from list.

PA 3: Write a program that accepts variable number of arguments and display the sum of all the elements present in it.

PA 4: Write a program to assign grades to students and display all the grades using keys() and get() method of dictionary.

PA 5: Write a program to count the occurrences of each element within a list.

PA 6: Write a program to print and store cubes of numbers into dictionary.

PA 7: Write a program to check if an element of a list is a prime number. If it is a prime, return True or return False.

PA 8: Write a function that accepts two positive integers viz. a and b and returns a list of all the even numbers between a and b.

PA 9: Write a program to pass a list to a function and return it in reverse order.

PA 10: Consider a list with five different Celsius values. Convert all the Celsius values into Fahrenheit.

References

Downey, Allen. *Think Python*. 2012. O'Reilly Media

Goodrich, Michael T., Roberto Tamassia, and Michael H. Goldwasser. 2013. *Data structures and algorithms in Python*. John Wiley & Sons.

Kachroo, Pushkin, and Kaan Özbay. 1999. Fuzzy Feedback Control for Dynamic Traffic Routing. In *Feedback control theory for dynamic traffic assignment*. Springer: 121–153.

McKinney, Wes. 2010. Data structures for statistical computing in python. In Proceedings of the 9th Python in Science Conference. vol. 445: 51–56.

Miller, Bradley N., and David L. Ranum. 2011. *Problem solving with algorithms and data structures using python*, Second Edition. Franklin, Beedle & Associates.

Srinivasa, K. G., G. M. Siddesh, and H. Srinidhi. 2018.*Network data analytics: A hands-on approach for application development*. Springer.

Walters, Gregory. 2014. Strings. In *The Python quick syntax reference*. Apress: 25–42.

7

Functions

7.1 Introduction

It is difficult to prepare and maintain a large-scale program and the identification of the flow of data subsequently becomes harder to understand. The best way of creating a programming application is to divide a big program into smaller subprograms and repeatedly call these modules. Using functions, an entire program can be divided into small, independent modules. This improves the code readability as well as the flow of execution as small modules can be managed easily (Zhang et al. 2014).

7.2 Types of Functions

Python supports the following functions:

7.2.1 Built-in Functions

Built-in functions are pre-defined functions of Python programming; you can call them to perform the task. For example: id(), type(), print(), etc.

7.2.2 User-Defined Functions

Users can define their own functions on the basis of their requirement. To define them, first define the functions as function definition. In a function definition, users have to declare a name for the new function and group of statements that will execute when the function will be called (Erickson 1975). For creating a user defined function, the mandatory keyword is 'def' and the optional keyword is 'return'.

Creating user-defined function:

```
def fun_name(parameters)
    ""documentstring""
    _____

    _____
    return data
```

DOI: 10.1201/9781003185505-7

Program: Write a function to print Hello three times in different lines.

```
def m1():
    print("Indian Army")
m1()
m1()
m1()
```

```
Indian Army
Indian Army
Indian Army
```

Example:

```
def m1(n):
    print(n,"Sir")
m1("Vijay")
m1("Vimal")
```

```
Vijay Sir
Vimal Sir
```

Program: Write a program that takes a number and displays its square as the output.

```
def m1(n):
    print("Square",n,":", n*n)
m1(3)
m1(7)
```

```
Square 3 : 9
Square 7 : 49
```

7.3 Return Statement

The return statement is used if you want to return some values from a function. It is also used to take control from the body of loop to the block where this function is called. The default return value will be None if you are not mentioning the return statement.

Program: Write a function that takes two numbers and displays their addition.

```
def add(a,b):
    return a+b
r=add(5,12)
print("Addition:",r)
print("Addition:",add(16,27))
```

```
Addition: 17
Addition: 43
```

Example:

```
def m1():
    print("Indian Army")
m1()
print(m1())
```

```
Indian Army
Indian Army
None
```

Program: Write a user-defined function to check whether the number is even or odd.

```
def evenodd(n):
    if n%2==0:
        print(n,":","Even Number")
    else:
        print(n,":","Odd Number")
evenodd(8)
evenodd(21)
```

```
8 : Even Number
21 : Odd Number
```

Program Write a user-defined function to identify the factorial of a given range of numbers.

```
def fact(n):
    r=1
    while n>=1:
        r=r*n
        n=n-1
    return r
for i in range(1,5):
    print("Factorial",i,":",fact(i))
```

```
Factorial 1 : 1
Factorial 2 : 2
Factorial 3 : 6
Factorial 4 : 24
```

Program: Write a program that calculates the addition and subtraction of two numbers.

```
def m1(a,b):
    sum=a+b
    sub=a-b
    return sum,sub
n1,n2=m1(75,68)
print("Addition :",n1)
print("Subtraction :",n2)
```

```
Addition : 143
Subtraction : 7
```

Program: Write a function that calculates the addition, subtraction, multiplication, and division of two numbers.

```
def calculator(a,b):
    sum=a+b
    sub=a-b
    mul=a*b
    div=a/b
    return sum,sub,mul,div
c=calculator(10,5)
print("Output:")
for i in c:
    print(i)
```

```
Output:
15
5
50
2.0
```

7.4 Arguments in a Function

Parameters are used to give inputs to a function. They are specified with a pair of parenthesis in the function's definition. When a programmer calls a function, the values are also passed to the function (Reeves 1991). In a function, there are two types of arguments, namely formal arguments and actual arguments. For example:

def m1(x,y):

----

m1(11,22)

Here, x and y behave as formal arguments whereas 11 and 22 behave as actual arguments. There are four types of arguments in Python.

7.4.1 Positional Arguments

Positional arguments are passed to the function at a proper positional order. The position and the number of the arguments should be compared. By changing the order, the result may change whereas by changing the number of arguments, an error message will be displayed.

Example:

```
def m1(n,m):
    print(n/m)
m1(20,5)
m1(5,20)
```

```
4.0
0.25
```

7.4.2 Keyword Arguments

If the programmer knows the parameter name used within the function then the parameter name can explicitly be used while calling the function. Here, the order of arguments is not mandatory, but the number of arguments should match.

Example:

```
def m1(n,m):
    print(n,m)
m1(n="Vijay",m="Author")
m1(m="Author",n="Vijay")

Vijay Author
Vijay Author
```

Example:

```
def m1(n,m):
    print(n,m)
m1("Vijay","Author")
m1("Vijay",m="Author")
#m1(n="Vijay","Author")   #invalid

Vijay Author
Vijay Author
```

The keyword and the positional arguments can be used together, but the order should be positional and then keyword argument (Hoffman and Shier 1980).

7.4.3 Default Arguments

Parameters within the function's definition may have the default values. You can assign the default value of a parameter with the help of assignment operator.

Example:

```
def m1(n="Swati"):
    print(n,"Author")
m1("Vijay")
m1()

Vijay Author
Swati Author
```

7.4.4 Variable Length Arguments

When you pass the distinct number of arguments each time, then they are referred to as variable length arguments. Internally, the defined values are taken in a tuple. This function can be called by declaring any number of arguments including the zero number.

Syntax: def p1 (*n):

Example:

```
def m1(*n):
    s=0
    for n1 in n:
        s=s+n1
    print("Additon=",s)
m1()
m1(10)
m1(2,3)
m1(3,6,8)

Additon= 0
Additon= 10
Additon= 5
Additon= 17
```

Note: The positional and the variable length arguments can be merged.

Example:

```
def m1(n1,*s):
    print(n1)
    for s1 in s:
        print(s1)
m1(10)
m1(30,50,60)
m1(1,"P",2,"Q")

10
30
50
60
1
P
2
Q
```

Once you are using variable length arguments, if you wish to take another argument then you must provide values as the keyword argument.

Example:

```
def m1(*s,n1):
    for s1 in s:
        print(s1)
    print(n1)
m1("P","Q",n1=10)

P
Q
10
```

Note: The keyword can be declared as variable length arguments by using **.

Example: def p1(**n):

This function may be used by declaring n number of the keyword arguments and all the n keywords are saved in the dictionary.

Example:

```
def m1(**n):
    for i,j in n.items():
        print(i,"=",j)
m1(a=7,b=4,c=6)
m1(Id=1,name="Vijay",exp=5)
```

```
a = 7
b = 4
c = 6
Id = 1
name = Vijay
exp = 5
```

Example:

```
def m1(a1,a2,a3=6,a4=9):
    print(a1,a2,a3,a4)
m1(13,12)
m1(16,22,36,41)
m1(30,34,a4=70)
m1(a4=1,a1=5,a2=8)
```

```
13 12 6 9
16 22 36 41
30 34 6 70
5 8 6 1
```

Note: In a program, the function is the combination of lines declared with a common name; the module is the combination of functions stored in a common file whereas the library is the combination of modules.

7.5 Scope of Variables

Python supports the following scope of variables.

7.5.1 Global Variables

Global variables are defined outside the functions; thus, they have global scope.

Program: Write a program to show the scope of global variables.

```
x=15
def m1():
    print(x)
def m2():
    print(x)
m1()
m2()
```

```
15
15
```

7.5.2 Local Variables

Variables and parameters that are initialized within the function including parameters are said to exist in that function's local scope. Variables that exist in the local scope are called the local variables.

Example:

```
def m1():
    x1=16
    print(x1)      # valid
def m2():
    print(x1)        #invalid
m1()
m2()
```

```
16
```

```
NameError: name 'x1' is not defined
```

7.6 Global Keyword

The global keyword is used:

1. To define the global variable inside the function
2. To declare the global variable available to function for modifications, if needed.

Example:

```
x2=30
def m1():
    x2=50
    print(x2)
def m2():
    print(x2)
m1()
m2()
```

```
50
30
```

In the above program, the x2 variable is declared before the function (acts as the global variable) as well as inside the function (acts as the local variable). When the x2 variable is called inside the m1() function (where x2 is already declared),the local variable gets priority and gets printed whereas when it is called inside the m2() function (where x2 is not declared), the global variable gets priority. If the global and the local variable share the similar name then the global variable can be accessed inside a function as:

Example:

```
x3=30
def m1():
     global x3
     x3=50
     print(x3)
def m2():
     print(x3)
m1()
m2()

50
50
```

7.7 Recursive Function

Python also supports the recursive feature, which means that a function is repeatedly calling itself. Thus, a function is said to be recursive if the statement declared in the body of the function calls itself, like calculating the factorial as follows is an example of recursive function.

$$fac(n) = n* fac(n-1)$$

Program: Write a program using recursion that calculates the factorial of a number.

```
def  fact(n):
     if n==0:
          r=1
     else:
          r=n*fact(n-1)
     return r
print("Factorial:",fact(6))
print("Factorial:",fact(8))

Factorial: 720
Factorial: 40320
```

7.8 Lambda Function

Lambda functions are named after the Greek letter lambda. They only have a code to execute that is associated with them.

Syntax: Name = lambda(variables): Code

A lambda function does not contain a return statement. It contains a single expression as a body and not a block of statements as a body.

When a function is declared without any name, such types of nameless function are regarded as anonymous or lambda functions. Its basic requirement is to use it instantly (Drechsler and Stadel 1987).

Program: Write a program that calculates the square of input number using the lambda function.

```
l=lambda n:n*n
print("Square  :",l(3))
print("Square  :",l(6))

Square  :  9
Square  :  36
```

Program: Write a program that calculates the addition of two numbers using the lambda function.

```
l=lambda a,b:a+b
print("Addition:",l(15,25))
print("Addition:",l(50,80))

Addition:  40
Addition:  130
```

Program: Write a program to identify the larger of two numbers using the lambda function.

```
l=lambda a,b:a if a>b else b
print("Biggest  :",l(12,22))
print("Biggest  :",l(30,50))

Biggest  :  22
Biggest  :  50
```

7.9 Filter() Function

In Python, the filter() function is used to filter the values of the specified sequence, on the basis of some prescribed condition.

Program: Write a program to print even numbers from a list by using the filter() function.

```
def m1(a):
    if a%2==0:
        return True
    else:
        return False
l=[0,6,11,17,30,28,33]
l1=list(filter(m1,l))
print(l1)

[0, 6, 30, 28]
```

Program: Write a program to distinguish even and odd numbers from the input list by using the filter() and lambda function.

```
l=[0,6,11,17,30,28,33]
l1=list(filter(lambda a:a%2==0,l))
print(l1)
l2=list(filter(lambda a:a%2!=0,l))
print(l2)
```

```
[0, 6, 30, 28]
[11, 17, 33]
```

7.10 Map() Function

In Python, the map() function takes a function and a list of all items as input and a new list is produced as an output that possesses the list of items given by the defined function for all the items.

Syntax: map(func,seq)

This function is used on all the elements of the sequence and produces a new sequence.

Example: (without lambda)

```
l=[11,22,33,44]
def m1(a):
    return 2*a
l1=list(map(m1,l))
print(l1)
```

```
[22, 44, 66, 88]
```

Example: (with lambda)

```
l=[11,22,33,44]
l1=list(map(lambda a:2*a,l))
print(l1)
```

```
[22, 44, 66, 88]
```

Program: Write a program using the lambda function that identifies the square of a number.

```
l=[11,22,33,44]
l1=list(map(lambda a:a*a,l))
print(l1)
```

```
[121, 484, 1089, 1936]
```

The map() function can be used on multiple lists having the same size also.

Syntax: map(lambda p,q: p*q,p1,p2)
             where, p is from p1 and q is from q2

Example:

```
l1=[11,22,33,44]
l2=[2,3,4,2]
l3=list(map(lambda x,y:x*y,l1,l2))
print(l3)
```

```
[22, 66, 132, 88]
```

7.11 Reduce() Function

The function reduce(func, seq) is continuously applied to the function func() of the sequence seq and produces a single value as output.

Example:

```
from functools import *
l=[11,22,33,44]
r=reduce(lambda a,b:a+b,l)
print(r)
r=reduce(lambda a,b:a*b,l)
print(r)
```

```
110
351384
```

Example:

```
from functools import *
r=reduce(lambda a,b:a+b,range(1,50))
print(r)
```

```
1225
```

In Python everything is considered as an object; internally, the functions are also considered as an object.

Example:

```
def m1():
    print("Indian Army")
print(m1)
print(id(m1))
```

```
<function m1 at 0x000001AAAD3F38B0>
1832562669744
```

7.12 Function Aliasing

Function aliasing is to give another name to an existing function.

Example:

```
def m1(n):
    print(n)
tmp=m1
print(id(m1))
print(id(tmp))
tmp('Vijay')
m1('Vijay')
```

```
1832562669312
1832562669312
Vijay
Vijay
```

After declaring a function, if it is deleted than that function can still be accessed by its alias name. In the following example, a function is created named as m1();its alias is created as tmp. When the function m1 is deleted, it can still be accessed by using its alias name.

Example:

```
def m1(n):
    print(n)
tmp=m1
tmp('Vijay')
m1('Vijay')
del m1
tmp('Swati')
m1('Parth')
```

```
Vijay
Vijay
Swati
```

```
NameError: name 'm1' is not defined
```

7.13 Nested Functions

The functions that are declared within the scope of another function are called nested functions. If this category of function definition is used, the inner function is the scope inside the outer function, so it is generally preferred when the inner function is returned or when it is being passed into another function.

Example:

```
def m1():
    print("m1() function started")
    def n1():
        print("n1() function execution")
    print("m1() function calling n1() function")
    n1()
m1()
n1()
```

```
m1() function started
m1() function calling n1() function
n1() function execution

NameError: name 'n1' is not defined
```

Example: (function returning other function)

```
def m1():
    print("m1() function started")
    def n1():
        print("n1() function execution")
    print("m1() function returning n1() function")
    return n1
m2=m1()
m2()
m2()
m2()
```

```
m1() function started
m1() function returning n1() function
n1() function execution
n1() function execution
n1() function execution
```

7.14 Decorator Functions

Decorator is an interesting attribute of Python, which is used to add new functionalities to the existing code. It is also referred to as meta-programming because during compilation time some parts of the program may try to change another part of the program. A decorator is any callable Python object that is used to change the function or the class (Slimick 1971). A reference to the function "func" or the class "C" is passed to a decorator and the decorator gives the modified function or class. The modified function or the classes usually contain the call to the original function "func" or the class "C".

Example:

```
def m1(n):
    print(n,"Welocome")
```

This is a function created that returns exactly the same output for distinct names.

On the contrary, if you wish to alter this function for providing a different message say, for the name Swati, it can be done without changing the m1() function by using the decorator.

Example: (On calling the m1() function, the décor function will be processed automatically)

```
def decor(fun):
    def m1(n):
        if n=="Swati":
            print("This is Swati")
        else:
            fun(n)
    return m1
@decor
def m2(n):
    print(n,"Welcome")
m2("Vijay")
m2("Vimal")
m2("Swati")
```

```
Vijay Welcome
Vimal Welcome
This is Swati
```

7.14.1 Calling of a Same Function Using and without Using a Decorator

Example:

```
def decor(fun):
    def m1(n):
        if n=="vimal":
            print("Hello Vimal")
        else:
            fun(n)
    return m1
def m2(n):
    print(n,"Welcome")
d=decor(m2)
m2("vijay")  #decorator wont be executed
m2("vimal")  #decorator wont be executed
d("vijay")   #decorator will be executed
d("vimal")   #decorator will be executed
```

```
vijay Welcome
vimal Welcome
vijay Welcome
Hello Vimal
```

Example:

```
def mul(fun):
    def m1(x,y):
        print("Var-1:",x,"Var-2:",y)
        if y==0:
            print("It's Zero")
            return
        else:
            return fun(x,y)
    return m1
@mul
def multiplication(x,y):
    return x*y
print(multiplication(14,5))
print(multiplication(14,0))
```

```
Var-1: 14 Var-2: 5
70
Var-1: 14 Var-2: 0
It's Zero
None
```

Example:

```
def Studentdecor(fun):
    def m1():
        print('Student Name')
        print('Subject')
        print('Package')
        fun()
    return m1
def m2():
    print('Company Placement')
Result=Studentdecor(m2)
Result()
```

```
Student Name
Subject
Package
Company Placement
```

7.14.2 Decorator Chaining

With the help of decorator chaining, you can declare various decorators for similar functions.

Example: @decor_1
             @decor
             def sum():

For the sum() function, you are using two decorator functions. At first, the inner decorator will get executed and then the outer decorator.

Example:

```
def decor1(fun):
    def m1():
        f=fun()
        return f*f
    return m1
def decor(fun):
    def m1():
        f=fun()
        return 3*f
    return m1
@decor1
@decor
def m2():
    return 5
print(m2())
```

```
225
```

7.15 Generator Functions

An important class of functions is the generators, which simplifies the job of writing the iterators. It is a routine that is used to manage the iteration behavior of a loop. It can be related to a function that produces an array. It comprises parameters, which are called and produce the sequence of numbers. When the generator functions are created, they will implement the iteration protocol automatically (Boehm and Demers 1986). Basically, the normal functions exit after returning the value whereas the generator function suspends automatically and then resumes the execution.

Example:

```
def method():
    yield 'P'
    yield 'Q'
    yield 'R'
m=method()
print(type(m))
print(next(m))
print(next(m))
print(next(m))
print(next(m))
```

```
<class 'generator'>
P
Q
R

StopIteration
```

Example:

```
def method(n):
    print("Numbers")
    while(n>0):
        yield n
        n=n-1
v=method(4)
for i in v:
    print(i)
```

```
Numbers
4
3
2
1
```

Example (to generate first n numbers):

```
def method(n1):
    n=1
    while n<=n1:
        yield n
        n=n+1
v=method(4)
for i in v:
    print(i)
```

```
1
2
3
4
```

Example:

```
v = method(5)
l1 = list(v)
print(l1)
```

```
[1, 2, 3, 4, 5]
```

Example (to generate Fibonacci numbers):

```
def fib():
    x,y=0,1
    while True:
        yield x
        x,y=y,x+y
for f in fib():
    if f>30:
        break
    print(f)
```

```
0
1
1
2
3
5
8
13
21
```

For normal collection:

```
l=[i*i for i in range(100000000000000000)]
print(l[0])
```

Here, you will get MemoryError as all the values are needed to be saved in memory.
For generators:

```
g=(i*i for i in range(100000000000000000000))
print(next(g))
```

```
0
```

7.16 Conclusion

In this chapter, the Python functions have been discussed in detail, starting with the importance of user-defined functions. This chapter covers creation, updating, and deletion of functions. It also covered the scope and lifetime of a variable. Later, the chapter covered all the variants of functions and concluded with the key distinction between decorator and generator functions.

Review Questions

1. What are keyword arguments?
2. What is the lambda function?

3. What are the characteristics of anonymous functions?
4. What do you mean by default arguments?
5. Briefly explain the function calling in Python.
6. What is a decorator?
7. Explain how classes can be used as decorators.
8. Differentiate between a function decorator and class decorator.

Programming Assignments

PA 1: Write a user-defined function factors(num) to calculate the factors of the given number.

PA 2: Write a program to define function dec_bin(num) to convert the existing decimal number into its equivalent binary number.

PA 3: Write a program to return multiple values from a function.

PA 4: Write a function calc_arithmetic(num1, num2) to calculate and return at once the result of arithmetic operations such as addition and subtraction.

PA 5: Write a program to access a local variable outside a function.

References

Boehm, Hans-Juergen and Alan Demers. 1986. Implementing Russell. *ACM SIGPLAN Notices* 21, no. 7: 186–195.

Drechsler, Karl-Heinz, and Manfred P. Stadel. 1987. The Pascal-XT code generator. *ACM SIGPLAN Notices* 22, no. 8: 57–78.

Erickson, David B. 1975. Array processing on an array processor. In Proceedings of the conference on Programming languages and compilers for parallel and vector machines: 17–24.

Hoffman, Karla L., and Douglas R. Shier. 1980: A test problem generator for discrete linear L 1 approximation problems. *ACM Transactions on Mathematical Software (TOMS)* 6, no. 4: 587–593.

Reeves, Alyson A. 1991. The worst order in not always the lexicographic order. *ACM SIGSAM Bulletin* 25, no. 4: 18–19.

Slimick, John. 1971. Current systems implementation languages: One user's view. In Proceedings of the SIGPLAN symposium on Languages for system implementation: 20–28.

Zhang, Rui, Jianzhong Qi, Martin Stradling, and Jin Huang. 2014. Towards a painless index for spatial objects. *ACM Transactions on Database Systems (TODS)* 39, no. 3: 1–42.

8

Modules

8.1 Introduction

Modules play a vital role in Python. In a module, you can bind randomly named attributes and can refer them. They are also used to divide the larger programs into smaller programs that are easy to manage and handle. A set of variables, functions, and classes saved to a common file is known as a module. The generally used functions can be defined in a module and imported, instead of copying their definitions in various programs. All the Python files with the extension .py act as the modules.

Example:

```
a = 100
def add(x,y):
    print("Addition:",x+y)
def sub(x,y):
    print("Subtraction:",x-y)
def mul(x,y):
    print("MUltiplication:",x*y)
def div(x,y):
    print("Division:",x/y)
```

If you want to use members of a module in the program then that module needs to be imported.

Syntax: import module_name/

The members can be accessed using the module name.

Syntax: module_name.variable
          module_name.func()

DOI: 10.1201/9781003185505-8

Example:

```
import Calculator
print(Calculator.a)
Calculator.add(100,200)
Calculator.sub(300,200)
Calculator.mul(100,200)
Calculator.div(400,200)
```

```
100
Addition: 300
Subtraction: 100
MUltiplication: 20000
Division: 2.0
```

When you are using the module in the program, the compiled file will be generated for the respective module that needs to be saved permanently in the hard disk (Kalicharan 2015).

8.2 Module Aliasing

The module can be renamed during importing it.

Example:

import Calculator as c
Here Calculator is an original module_name and c is its alias name.

The members can be accessed by alias name also.

Example:

```
import Calculator as c
print(c.a)
Calculator.mul(100,200)
Calculator.div(400,200)
```

```
100
MUltiplication: 20000
Division: 2.0
```

Some particular members of the module can be imported by from .... import.
The benefit is that the members can be accessed directly without using the module name.

Example:

```
from Calculator import a,mul
print(a)
mul(100,200)
div(400,200)
add(200,50)
```

```
100
MUltiplication: 20000
Division: 2.0
Addition: 250
```

Example: (to import all the members of a module)

```
from Calculator import *
print(a)
add(60,30)
div(50,2)
```

```
100
Addition:  90
Division:  25.0
```

Different ways to import a module:

- import module_name
- import module_1,module_2,module_3
- import module_1 as p
- import module_1 as p,module_2 as q,module_3
- from module import membr
- from module import membr_1,membr_2,membr_3
- from module import membr_1 as p
- from module im

8.3 Member Aliasing

Example:

```
from Calculator import a as b,add as sum
print(b)
sum(50,80)
```

```
100
Addition:  130
```

After defining the alias name, the original name should not be used.

Example:

```
from Calculator import a as b,add as sum
print(b)
sum(50,80)
```

```
100
Addition:  130
```

8.4 Reloading a Module

Even after importing the module multiple times, it gets loaded only once.

Example:

```
import Parth
import Parth
import Parth
import Parth
print("My Module is define")
```

```
This Parth Modules for python
My Module is define
```

In the above defined program, the module named Parth gets loaded only once, even after importing it multiple times. But the problem here is, after loading the Parth module, it gets updated out-side and the updated version of the defined module will not be available to the program. This problem can be resolved by reloading the module explicitly as needed. It can be reloaded using the reload() function of the imp module (Van Rossum and Drake 2011).

Example:

```
import importlib
importlib.reload(Parth)
```

```
<module 'Parth' from 'C:\\Users\\lenovo\\Parth\\__init__.py'>
```

Example:

```
import imp
import Parth
import Parth
from imp import reload
imp.reload(Parth)
imp.reload(Parth)
imp.reload(Parth)
print("My Module is define")
```

```
This Parth Modules for python
This Parth Modules for python
This Parth Modules for python
This Parth Modules for python
My Module is define
```

8.5 Dir() Function

In Python, there is an in-built function, dir(), which is used to list all the available members of the specific module.

Example:

```
a=10
b=20
def m1():
    print("Vijay")
print(dir())
```

```
['In', 'Out', '_', '__', '___', '__builtin__',

'__builtins__', '__doc__', '__loader__',

'__name__', '__package__', '__spec__',

'_dh', '_i', '_i1', '_ih', '_ii', '_iii', '_oh'

'a', 'b', 'exit', 'get_ipython', 'm1', 'quit']
```

Example: (displaying members of a specified module)

```
a = 100
def add(x,y):
    print("Addition:",x+y)
def sub(x,y):
    print("Subtraction:",x-y)
def mul(x,y):
    print("MUltiplication:",x*y)
def div(x,y):
    print("Division:",x/y)
```

Example:

```
import Calculator
print(dir(Calculator))
```

```
['__builtins__', '__cached__', '__doc__', '__file__',

'__loader__', '__name__', '__package__',

'__path__', '__spec__', 'a', 'ad

d', 'cal', 'div', 'mul', 'sub']
```

During execution, for all the modules the Python interpreter automatically appends a few special properties for the internal usage. On the basis of requirement, these properties can also be accessed in the program (Kalicharan 2008).

Example:

```
print(__builtins__)
print(__cached__)
print(__doc__)
print(__file__)
print(__loader__)
print(__name__)
print(__package__)
print(__spec__)
```

```
<module 'builtins' (built-in)>
```

```
NameError: name '__cached__' is not defined
```

Special variables _name_

In Python, internally for all the programs a special variable _name_ gets added. It includes details of whether the program is executed individually or executed as a module. If the program is executed individually then the value of the variable will be _main_ whereas if it is executed as a module then the value of the variable will be the name of module where it is declared.

Example:

```
def m1():
    if __name__=='__main__':
        print("Run program part")
    else:
        print("Other program rus as a module")
m1()
```

```
Run program part
```

Example: (m1() function is imported using P_Face module)

```
import P_Face
P_Face.m1()
```

```
Other program rus as a module
Other program rus as a module
```

8.6 Math Module

A module is a combination of variables, functions, or classes, etc. It is a module that supports various functions for performing various mathematical operations.

Importing math module:

- import math

After importing this module, any function can be called.
   Creating alias name:

- import math as m

After creating an alias name, functions and variables can be accessed.

Example:

```
import math
print(math.sqrt(16))
print(math.pi)
```

```
4.0
3.141592653589793
```

Importing a specific member of a module explicitly:

- from math import sqrt
- from math import sqrt,pi

Example:

```
from math import sqrt, pi
print(sqrt(16))
print(pi)
print
```

```
4.0
3.141592653589793

<function print>
```

The math module has various functions:

- sqrt(x)
- floor(x)
- ceil(x)
- log(x)
- sin(x)
- cos(x)
- tan(x) and many more.

Example:

```
from math import *
print(sqrt(4))
print(ceil(10.1))
print(floor(10.1))
print(fabs(-10.6))
print(fabs(10.6))
```

```
2.0
11
10
10.6
10.6
```

Help can be taken for any module using the help() function.

Example:

```
import math
help(math)
```

```
Help on built-in module math:

NAME
    math

DESCRIPTION
    This module provides access to the mathematical functions
    defined by the C standard.

FUNCTIONS
    acos(x, /)
        Return the arc cosine (measured in radians) of x.

---------------------
---------------------
---------------------
    trunc(x, /)
        Truncates the Real x to the nearest Integral toward 0.

        Uses the __trunc__ magic method.

DATA
    e = 2.718281828459045
    inf = inf
    nan = nan
    pi = 3.141592653589793
    tau = 6.283185307179586

FILE
    (built-in)
```

8.7 Random Module

The random module is used to generate various random numbers. It comprises several functions that are defined below:

1. random() Function: It produces any float value in the range 0 to 1, without inclusive 0 and 1.

Example:

```
from random import *
for v in range(6):
    print(random())
```

```
0.026449297903502034
0.5626551544033295
0.06275088939132678
0.8857573006432503
0.08862585817459445
0.45969071713949683
```

2. randint() Function: It produces any random integer between specified numbers, with including them.

Example:

```
from random import *
for v in range(5):
    print(randint(1,10))
```

```
3
10
9
10
8
```

3. uniform() Function: It produces any random float values between specified numbers, without including them.

Example:

```
from random import *
for v in range(5):
    print(uniform(1,6))
```

```
4.810383860183087
4.521809122095024
1.6915175123295803
3.7204545754267
4.330677523045923
```

4. randrange ([begin], end, [step]): It produces a random number from the defined range. Here, the begin argument is optional and if not defined, then the by-default value is 0. The step argument is also optional and if not defined then, the by-default value is 1.

Example:

```
from random import *
for v in range(5):
    print(randrange(8))
```

```
1
5
3
4
3
```

Example:

```
from random import *
for v in range(5):
    print(randrange(1,8))
```

```
5
3
2
7
6
```

Example:

```
from random import *
for v in range(5):
    print(randrange(1,10,2))
```

```
1
7
1
3
1
```

5. choice() Function: It gives a random object from the specified list or the tuple.

Example:

```
from random import *
l=["Vijay","Shashwat","Parth","Vimal","Swati"]
for v in range(6):
    print(choice(l))
```

```
Vijay
Vijay
Swati
Parth
Shashwat
Vimal
```

8.8 Packages

A package follows the concept of encapsulation to group interrelated modules as a single unit. It acts like a file, a folder, or a directory representing a group of Python modules. It can contain sub-packages also. Any directory or any folder containing _init_.py file, is treated as a Python package. It may be empty (Campbell 2014). Naming conflict can be resolved using packages, as the components can be identified uniquely.

Example:

```
import Pack1.prog1
Pack1.prog1.fun()
```

```
This is fun function
```

Example:

```
from Pack2.prog1 import fun1
from Pack2.Pack3.prog2 import fun2
fun1()
fun2()
```

```
This is fun1 function
This is func2 function
```

8.9 Conclusion

In this chapter, you have learned methods to create and import modules in Python. Also, you have seen variant techniques to import and use custom and in-built modules in Python, creation of executable module as a stand-alone script, and organization of modules into packages and sub-packages.

Review Questions

1. Explain the importance of the import statement.
2. Explain the usage of reload() function.
3. How is a module located in Python?
4. Explain the uniform() function.

Programming Assignments

PA 1: Write a program to get days between two dates.
PA2: Write a program to get current time in milliseconds.
PA 3: Write a program to subtract five days from the current date.
PA 4: Write a program to get the current time.
PA 5: Write a program to convert a string to datetime.

References

Campbell, Matthew. 2014. *Objective – C quick syntax reference*. Apress.
Kalicharan, Noel. 2008. *C programming – A beginner's course*. Create Space.
Kalicharan, Noel. 2015. *Learn to program with C*. Apress.
Van Rossum, Guido, and Fred L. Drake. 2011. *The Python language reference manual*. Network Theory.

9

Basic Concepts of Object-Oriented Programming

9.1 Introduction

Object oriented programming (OOPs)is a powerful concept for creating programs. So, far we have discussed the procedural concepts in programming. Before going to object oriented programming, we need to understand the basic terminologies used in object oriented programming. While procedure oriented programming gives importance to functions, object oriented programming focuses on objects. The basic concepts related to OOPs are classes, objects, encapsulation, inheritance, and polymorphism.

9.2 Class

The focal point in object oriented programming is the class and object. The classes are the building block of Python. The class constructs a new type, and the objects are the instances of the class. The class uses the template or the blueprint with which the objects are created. A new local namespace is being created by the class wherein all the attributes are defined (Cahill and Lafferty 2012). They may be functions or data.

Syntax to define a class: class class_name:
                                 '''document string'''

A document string shows the description of a class. The string is optional in the doc class. The doc string can be accessed by the following techniques:

1. print(class_name.__doc__)
2. help(class_name)

DOI: 10.1201/9781003185505-9

Program: Create a simple class in Python.

```
class A:
        ''''Document Information'''
print(A.__doc__)
help(A)
```

```
''Document Information
Help on class A in module __main__:

class A(builtins.object)
 |    ''Document Information
 |
 |    Data descriptors defined here:
 |
 |    __dict__
 |        dictionary for instance variables (if defined)
 |
 |    __weakref__
 |        list of weak references to the object (if defined)
```

Program: Create a simple Python class and an object of the class.

```
class Python:
    ''''' Python programming approach'''
    def __init__(self):
        self.n="Abhishek"
        self.r=40
    def m1(self):
        print("Name:",self.n)
        print("Size:",self.r)
p=Python()
print(p.m1())
```

```
Name: Abhishek
Size: 40
None
```

9.3 Object

An object is the physical existence of a class. The 'n' number of objects can be created for a class. The reference variable is the one that is used to refer to the object, methods, and its properties.

Syntax to create object: ref_variable = class_name()

Example: p = Stdnt()

Example:

```
class A:
    def __init__(self,n):
        self.n=n
    def m1(self):
        print("Name:",self.n)
a=A("Shashwat")
a.m1()
```

```
Name: Shashwat
```

9.4 Self Variable

It is a default variable pointing to the current object. By using it, instance variables and instance methods can be accessed.

1. Inside the constructor, the first parameter should be self.
   Example: def _init_self():
2. Inside instance methods, the first parameter should be self.
   Example: def pqr(self):

9.5 Constructor

The _init_() method has a special significance in Python classes. The _init_() method is automatically executed when an object of a class is created. The method is useful to declare and initialize the instance variables. The constructor is only executed once per object and at least one argument has to be provided to it (Lee 2013). If the user doesn't provide the constructor then the PVM will provide the default constructor.

Program: Write a program to demonstrate that the constructor will execute only once per object.

```
class A:
    def __init__(self):
        print("Constructor")
    def m1(self):
        print("Method")
a1=A()
a1.m1()
```

```
Constructor
Method
```

Program: Write some Python code using the constructor.

```python
class A:
    ''''' Information here'''
    def __init__(self,a,b):
        self.n=a
        self.r=b
    def m1(self):
        print("Name:{}\nId:{}".format(self.n,self.r))
a1=A("Vimal",1020)
a1.m1()
```

```
Name:Vimal
Id:1020
```

The differences between methods and constructors are tabulated in Table 9.1.

TABLE 9.1
Difference between method and constructor

Method	Constructor
Any name can be used to declare a method.	The constructor is always declared using _init_
The method is executed on calling it.	The constructor is executed automatically when the object gets created.
It may be called any number of times, for each object.	It is called only once, for each object.
The business logic can be written inside methods.	The instance variables are declared and initialized in it.

9.6 Types of Variables

9.6.1 Instance Variables

Instance variables are the ones whose value varies from object-to-object. For each object, a distinct copy of the instance variable will be produced. They are also known as object level variables.

9.6.1.1 *Declaring Instance Variables*

a) In the constructor by self keyword

The instance variables can be declared in the constructor using the self variable. When the objects are created, these variables get appended to the object automatically.

Example:

```python
class A:
    def __init__(self):
        self.name='Parth'
        self.ids=2020
a=A()
print(a.__dict__)
```

```
{'name': 'Parth', 'ids': 2020}
```

b) In the instance method by self keyword.

Instance variables can be defined inside the instance method using the self keyword. If the instance variable is defined in the instance method, that instance variable will be appended once the method is called.

Example:

```
class A:
    def __init__(self):
        self.a=100
    def m1(self):
        self.b=300
a1=A()
a1.m1()
print(a1.__dict__)
```

```
{'a': 100, 'b': 300}
```

c) Outside the class by object reference variable.

The instance variables can be added outside class of a specific object.

Example:

```
class A:
    def __init__(self):
        self.a=10
    def m1(self):
        self.b=30
a1=A()
a1.m1()
a1.c=40
print(a1.__dict__)
```

```
{'a': 10, 'b': 30, 'c': 40}
```

9.6.1.2 *Accessing Instance Variables*

Instance variables can be accessed within the class by the self variable and outside the class by object reference.

Example:

```
class A:
    def __init__(self):
        self.a=3
        self.b=5
    def m1(self):
        print(self.a)
        print(self.b)
a1=A()
a1.m1()
print(a1.a,a1.b)
```

```
3
5
3 5
```

9.6.1.3 Deleting Instance Variables

a) Instance variables can be deleted within the class as follows:
    del self.variable_name
b) Instance variables can be deleted outside the class as follows:
    del objectreference.variable_name

Program: Create a class to show the deletion of an instance variable from the object.

```
class A:
    def __init__(self):
        self.p=3
        self.q=5
        self.r=7
    def m1(self):
        del self.q
a1=A()
print(a1.__dict__)
a1.m1()
print(a1.__dict__)
del a1.r
print(a1.__dict__)
```

```
{'p': 3, 'q': 5, 'r': 7}
{'p': 3, 'r': 7}
{'p': 3}
```

Example:

```
class A:
    def __init__(self):
        self.p=13
        self.q=25
        self.r=38
a1=A()
a2=A()
del a1.p
print(a1.__dict__)
print(a2.__dict__)
```

```
{'q': 25, 'r': 38}
{'p': 13, 'q': 25, 'r': 38}
```

If you modify the values of an instance variable of an object then these modifications will not be affected by the remaining objects, as an individual copy of instance variables is there.

Example:

```
class A:
    def __init__(self):
        self.p=3
        self.q=5
a1=A()
a1.p=10
a1.q=20
a2=A()
print('a1:',a1.p,a1.q)
print('a2:',a2.p,a2.q)
```

```
a1: 10 20
a2: 3 5
```

9.6.2 Static Variables

Static variables are used if the value of variables does not change from object-to-object, this category of variables has to be defined in the class directly but outside of the method. For the complete class, only a single copy of the static variable will be constructed and shared by all the objects of the corresponding class. The static variables can be accessed by class name or by object reference. They are also known as class level variables.

Example:

```
class A:
    p=12
    def __init__(self):
        self.q=22
a1=A()
a2=A()
print('a1:',a1.p,a1.q)
print('a2:',a2.p,a2.q)
A.p=50
a1.q=60
print('a1:',a1.p,a1.q)
print('a2:',a2.p,a2.q)
```

```
a1: 12 22
a2: 12 22
a1: 50 60
a2: 50 22
```

9.6.2.1 *Declaration of Static Variables*

The static variables are declared inside the class and outside the method. It can be even declared within the constructor by class name or within the instance method by class name. It can also be declared in the class method by using the class name or by using the cls variable (Van and Drake 2003). Moreover, it can also be declared in the static method by using the class name.

Example:

```
class A:
    p=2
    def __init__(self):
        A.q=3
    def p1(self):
        A.r=4
    @classmethod
    def p2(cls):
        cls.s=5
        A.t=6
    @staticmethod
    def p3():
        A.u=7
print(A.__dict__)
a=A()
print(A.__dict__)
a.p1()
print(A.__dict__)
A.p2()
print(A.__dict__)
A.p3()
print(A.__dict__)
A.v=9
print(A.__dict__)
```

```
{'__module__': '__main__', 'p': 2, '__init__': <function A.__
38A6B80>, 'p2': <classmethod object at 0x000001FE537957F0>, '
ribute '__dict__' of 'A' objects>, '__weakref__': <attribute
's': 5, 't': 6, 'u': 7}
{'__module__': '__main__', 'p': 2, '__init__': <function A.__
38A6B80>, 'p2': <classmethod object at 0x000001FE537957F0>, '
ribute '__dict__' of 'A' objects>, '__weakref__': <attribute
's': 5, 't': 6, 'u': 7, 'v': 9}
```

9.6.2.2 *Accessing Static Variables*

1. Within the constructor: by self or class_name
2. Within the instance method: by self or class_name
3. Within the class method: by cls variable or class_name
4. Within the static method: by class_name
5. From outside of class: by object reference or class_name

Example:

```
class A:
    p=50
    def __init__(self):
        print(self.p)
        print(A.p)
    def n1(self):
        print(self.p)
        print(A.p)
    @classmethod
    def n2(cls):
        print(cls.p)
        print(A.p)
    @staticmethod
    def n3():
        print(A.p)
a=A()
print(A.p)
print(a.p)
a.n1()
a.n2()
a.n3()

50
50
50
50
50
50
50
50
50
```

The value of static variables can be modified inside or outside the class by class_name, but within the class method, by cls variable.

Example:

```
class A:
    p=50
    @classmethod
    def n1(cls):
        cls.p=100
    @staticmethod
    def n2():
        A.p=200
print(A.p)
A.n1()
print(A.p)
A.n2()
print(A.p)

50
100
200
```

Example:

```
class A:
    p=50
    def n1(self):
        self.p=100
a1=A()
a1.n1()
print(A.p)
print(a1.p)
```

```
50
100
```

Example:

```
class A:
    p=50
    def __init__(self):
        self.q=80
a1=A()
a2=A()
print('a1:',a1.p,a1.q)
print('a2:',a2.p,a2.q)
a1.p=150
a1.q=250
print('a1:',a1.p,a1.q)
print('a2:',a2.p,a2.q)
```

```
a1:  50 80
a2:  50 80
a1:  150 250
a2:  50 80
```

Example:

```
class A:
    p=50
    def __init__(self):
        self.q=80
a1=A()
a2=A()
A.p=150
a1.q=250
print(a1.p,a1.q)
print(a2.p,a2.q)
```

```
150 250
150 80
```

Example:

```
class A:
    p=50
    def __init__(self):
        self.q=80
    def n1(self):
        self.p=150
        self.q=250
a1=A()
a2=A()
a1.n1()
print(a1.p,a1.q)
print(a2.p,a2.q)
```

```
150 250
50 80
```

Example:

```
class A:
    p=50
    def __init__(self):
        self.q=80
    @classmethod
    def n1(cls):
        cls.p=150
        cls.q=250
a1=A()
a2=A()
a1.n1()
print(a1.p,a1.q)
print(a2.p,a2.q)
print(A.p,A.q)
```

```
150 80
150 80
150 250
```

9.6.2.3 Deleting Static Variables

a) The static variables can be deleted from anywhere as follows:

   syntax: del class_name.variable_name

b) The cls variable can also be used within the class method

   del cls.variable_name

Example:

```
class A:
    p=50
    @classmethod
    def n1(cls):
        del cls.p
A.n1()
print(A.__dict__)
```

```
{'__module__': '__main__', 'n1': <classmethod object at 0x00000206B1D83D08>
'__weakref__': <attribute '__weakref__' of 'A' objects>, '__doc__': None}
```

Example:

```
class A:
        p=50
        def __init__(self):
            A.q=80
            del A.p
        def m1(self):
            A.r=90
            del A.q
        @classmethod
        def m2(cls):
            cls.s=40
            del A.r
        @staticmethod
        def m3():
            A.t=50
            del A.s
print(A.__dict__)
a=A()
print(A.__dict__)
a.m1()
print(A.__dict__)
a.m2()
print(A.__dict__)
a.m3()
print(A.__dict__)
a.u=60

print(A.__dict__)
del a.r
print(A.__dict__)
del Test.t
print(Test.__dict__)
```

```
{'__module__': '__main__', 'p': 50, '__init__': <function A..
538D0B80>, 'm2': <classmethod object at 0x000001FE538A9D30>,
tribute '__dict__' of 'A' objects>, '__weakref__': <attribut
{'__module__': '__main__', '__init__': <function A.__init__
>, 'm2': <classmethod object at 0x000001FE538A9D30>, 'm3': <
'__dict__' of 'A' objects>, '__weakref__': <attribute '__wea
{'__module__': '__main__', '__init__': <function A.__init__
>, 'm2': <classmethod object at 0x000001FE538A9D30>, 'm3': <
'__dict__' of 'A' objects>, '__weakref__': <attribute '__wea
{'__module__': '__main__', '__init__': <function A.__init__
>, 'm2': <classmethod object at 0x000001FE538A9D30>, 'm3': <
'__dict__' of 'A' objects>, '__weakref__': <attribute '__wea
{'__module__': '__main__', '__init__': <function A.__init__
>, 'm2': <classmethod object at 0x000001FE538A9D30>, 'm3': <
'__dict__' of 'A' objects>, '__weakref__': <attribute '__wea
{'__module__': '__main__', '__init__': <function A.__init__
>, 'm2': <classmethod object at 0x000001FE538A9D30>, 'm3': <
'__dict__' of 'A' objects>, '__weakref__': <attribute '__wea
```

```
AttributeError: r
```

Example:

```
class A:
    p=5
a1=A()
del a1.p
```

```
AttributeError: p
```

The static variables can be deleted or modified by the class_name or the cls variable.

Example:

```
import sys
class Acc_Holder:
    bank='SBI'
    def __init__(self,n,bal=0.0):
        self.n=n
        self.bal=bal
    def deposit(self,amt):
        self.bal=self.bal+amt
        print('Balance:',self.bal)
    def withdraw(self,amt):
        if amt>self.bal:
            print('Insufficient Funds')
            sys.exit(0)
        self.bal=self.bal-amt
        print('Balance:',self.bal)
print(Acc_Holder.bank)
n=input('Enter Name:')
ah=Acc_Holder(n)
while True:
    print('d-Deposit \nw-Withdraw \ne-exit')
    o=input('Choose your option:')
    if o=='d' or o=='D':
        amt=float(input('Enter amount:'))
        ah.deposit(amt)
    elif o=='w' or o=='W':
        amt=float(input('Enter amount:'))
        ah.withdraw(amt)
    elif o=='e' or o=='E':

        print('Thanks for Banking')
        sys.exit(0)
    else:
        print('Try Again This is not correct')
```

```
SBI
Enter Name:v
d-Deposit
w-Withdraw
e-exit
Choose your option:400
Try Again This is not correct
d-Deposit
w-Withdraw
e-exit
Choose your option:d
Enter amount:400
Balance: 400.0
d-Deposit
w-Withdraw
e-exit
Choose your option:w
Enter amount:200
Balance: 200.0
d-Deposit
w-Withdraw
e-exit
Choose your option:e
Thanks for Banking
```

9.6.3 Local Variables

You can define the variables directly within the method by using the local or temporary variables. They are declared at the time of method processing and destroyed when the method is completed. They are also known as method level variables (Pilgrim and Willison 2009).

Program: Write a class to show the usage of method level variable.

```
class A:
    def method1(self):
        p=10
        print(p)
    def method2(self):
        q=20
        print(q)
a=A()
a.method1()
a.method2()

10
20
```

9.7 Types of Methods

In Python, there are three types of methods, discussed below:

9.7.1 Instance Methods

Inside the method implementation, if you are using instance variables these methods are known as instance methods. The self variable is declared within the instance method.

def p1(self):

By using the self variable within the method, you will be able to access the instance variables. Within the class, the self variable is used to call the instance method whereas outside the class, the object reference is used to call the instance method.

Example:

```python
class Emp:
    def __init__(self,emp_n,emp_sal):
            self.emp_n=emp_n
            self.emp_sal=emp_sal
    def method1(self):
        print(self.emp_n)
        print('Salary:',self.emp_sal)
    def method2(self):
        if self.emp_sal>=80000:
            print('Grade Pay 8')
        elif self.emp_sal>=70000:
            print('Grade Pay 7')
        elif self.emp_sal>=60000:
            print('Grade Pay 6')
        else:
            print('CONSOLIDATED ')
n=int(input('Enter Number of Emp:'))
for i in range(n):
    emp_n=input('Enter Name:')
    emp_sal=int(input('Enter Salary:'))
    e= Emp(emp_n,emp_sal)
    e.method1()
    e.method2()
    print()
```

```
Enter Number of Emp:1
Enter Name:vijay
Enter Salary:75000
vijay
Salary: 75000
Grade Pay 7
```

9.7.2 Class Methods

Inside the method implementation, if you are using the class variables then these methods should be declared as the class method. The class method can be declared explicitly by @classmethod decorator. One should provide the cls variable for the class method during declaration. The class method can be called by class_name or the object reference variable.

Program: Create a class showing the significance of the class method.

```python
class Book:
    authors=4
    @classmethod
    def writing(cls,name):
        print('{} writing book with {} authors'.
            format(name,cls. authors))
Book.writing('Vijay')
```

```
Vijay writing book with 4 authors
```

Program: Create a class to track the number of objects created for a class.

```python
class A:
    count=0
    def __init__(self):
        A.count =A.count+1
    @classmethod
    def method(cls):
        print('Number of objects created:',cls.count)
a1=A()
a2=A()
A.method()
a3=A()
a4=A()
A.method()
```

```
Number of objects created: 2
Number of objects created: 4
```

9.7.3 Static Methods

Static methods are the general utility methods. In the static methods, no instance or class variables are used. In this, self or cls arguments are not provided during declaration. The static method is declared explicitly by the @staticmethod decorator. The static methods are accessed by class_name or the object reference.

Program: Write a program to show the significance of static method.

```python
class Shashwat:
    @staticmethod
    def add(a,b):
        print('Addition:',a+b)
    @staticmethod
    def mul(a,b):
        print('Multiplication:',a*b)
    @staticmethod
    def avg(a,b):
        print('Average:',(a+b)/2)
Shashwat.add(5,8)
Shashwat.mul(25,5)
Shashwat.avg(30,60)
```

```
Addition: 13
Multiplication: 125
Average: 45.0
```

9.8 Setter and Getter Methods

The getter and setter methods are used in many object oriented programming languages to provide data encapsulation. They are also known as mutator methods. The getter method is used for retrieving data whereas the setter method is used to set a new value for the data

(Solis and Schrotenboer 2018). In OOP languages, the attributes of a class are made private to hide and protect them from other code.

Syntax of setter method:
```
def setVariable (self, variable):
    self.variable = variable
```

Syntax of getter method:
```
def getVariable(self):
    return self.variable
```

Example:

```
class Emp:
    def setemp_n(self,emp_n):
        self.emp_n=emp_n
    def getemp_n(self):
        return self.emp_n
    def setemp_sal(self,emp_sal):
        self.emp_sal=emp_sal
    def getemp_sal(self):
        return self.emp_sal
n1=int(input('Enter number of Emp:'))
for i in range(n1):
    e=Emp()
    emp_n=input('Enter Name:')
    e.setemp_n( emp_n)
    emp_sal=int(input('Enter Salary:'))
    e.setemp_sal(emp_sal)
    print(e.getemp_n())
    print('Salary:',e.getemp_sal())
    print()
```

```
Enter number of Emp:1
Enter Name:dd
Enter Salary:3000
dd
Salary: 3000
```

9.9 Passing Members of One Class to Other Class

The members of one class can be accessed within another class.

Example:

```
class Book:
    def __init__(self,name,pages,price):
        self.name=name
        self.pages=pages
        self.price=price
    def method(self):
        print('Book Name:',self.name)
        print('Book pages:',self.pages)
        print('Book Price:',self.price)
class update:
    def method1(b):
        b.price=b.price+6000
        b.method()
b1=Book("Python",350,4000)
update.method1(b1)
```

```
Book Name: Python
Book pages: 350
Book Price: 10000
```

9.10 Conclusion

In the earlier chapters, we have discussed the procedural concepts of programming and, after studying this chapter, you must have polished your basics of object oriented programming. Objects, classes, and methods are the key concepts that are used here.

Review Questions

1. What is a class?
2. How are attributes added to a class?
3. What is a self-parameter?
4. What can be done with overriding a method?
5. List the applications of a special class attribute.

Programming Assignments

PA 1: Create a class to convert an integer to a Roman numeral.
PA 2: Create a class to convert a Roman numeral to an integer.
PA 3: Create a class to find the area of a circle.
PA 4: Create a class to find the perimeter of a rectangle.

References

Cahill, Vinny, and Donal Lafferty. 2012. *Learning to program the object-oriented way with C.* Springer Science & Business Media.

Lee, Keith. 2013. *Pro Objective-C.* Apress.

Lutz, Mark. 2013. *Learning Python: Powerful object-oriented programming.* O'Reilly Media.

Naidu, C. 2020. Language fundamentals on Python. Notes of Python programming language document by Durga Software Solutions uploaded as open source document. www.scribd.com/document/427237165/Python-notes (accessed January 3, 2021).

Phillips, Dusty. 2010. *Python 3 object oriented programming.* Packt Publishing.

Pilgrim, Mark, and Simon Willison. 2009. *Dive into Python 3.* Vol. 2. Apress.

Solis, Daniel, and Cal Schrotenboer. 2018. *Illustrated C# 7: The C# language presented clearly, concisely, and visually.* Apress.

Van Rossum, Guido, and Fred L. Drake. 2003. *Python language reference manual.* Network Theory.

10

Advanced Concepts of Object-Oriented Programming

10.1 Introduction

In this chapter, we will move to advanced concepts in detail like abstraction, polymorphism, inheritance, and data encapsulation, which adds an extra layer of security to the developed program using the object oriented approach. Data abstraction provides safety and security to the data. Polymorphism allows the same interface for different objects, so programmers can write code more efficiently. These concepts are very useful while solving complex problems. As you increase your coding skills and learn more such advanced topics, you will be able to find out where and when to apply the concepts of object oriented programming will be helpful. The idiosyncrasy of object oriented programming is seen by the knowledge that most of the modern programming languages are either fully object oriented or support object oriented programming.

10.2 Inner Class

The declaration of a class inside another class is known as an inner class. They are preferred when no object type exists and there is no chance of another type of object.

Example: Without the existence of a parent object, the child object cannot exist.

class parent:
_____

class child:
_____

For the existence of an inner class object, the outer class object must exist. Thus, the inner class object should be linked with an outer class object.

DOI: 10.1201/9781003185505-10

Program: Write a program to show the association of an inner class with an outer class.

```python
class External:
    def __init__(self):
        print("Outer Class")
    class Internal:
        def __init__(self):
            print("Inner Class")
        def method(self):
            print("Inner class Method")
o=External()
i=o.Internal()
i.method()
```

```
Outer Class
Inner Class
Inner class Method
```

Ways to call the inner class method:

a) o = Outer(). o = Outer()

   i = o.Inner()

   i.p1()

b) i = Outer(). Inner()

   i.p1()

c) Outer().Inner().p1()

Example:

```python
class Scientist:
    def __init__(self):
        self.name='Dr. Kalam'
        self.info=self.Info()
    def method(self):
        print('Name:',self.name)
    class Info:
        def __init__(self):
            self.height=5.7
            self.weight=70
            self.color="Dark"
        def method(self):
            print('Height:{} Weight:{} Color:{}'.
                  format(self.height,self.weight,self.color))
s=Scientist()
s.method()
x=s.info
x.method()
```

```
Name: Dr. Kalam
Height:5.7 Weight:70 Color:Dark
```

Program: Write a program by declaring any number of inner classes inside the main class.

```python
class Company:
    def __init__(self):
        self.name = 'ABCD'
        self.doe = self.DOE()
        self.type = self.Type()
    def method(self):
        print(self.name)
    class DOE:
        def m1(self):
            print('m1 method')
    class Type:
        def m2(self):
            print('m2 method')
c=Company()
c.method()
c.doe.m1()
c.type.m2()
```

```
ABCD
m1 method
m2 method
```

10.3 Garbage Collection

Python performs automatic garbage collection. This means that it deletes all the objects automatically that are no longer needed and that have gone out of scope to free the memory space. The process by which Python periodically reclaims unwanted memory is known as garbage collection. Python's garbage collector runs in the background during program execution. It immediately takes action as soon as an object's reference count reaches zero (Alchin 2010). By default, the garbage collector is in the enable mode, but it can be disabled on the basis of requirement.

Functions of the gc module:

- gc.isenabled() : Returns True if GC enabled
- gc.disable() : To disable GC explicitly
- gc.enable() : To enable GC explicitly

Program: Write a program to check if the garbage collector is enabled or disabled.

```python
import gc
print(gc.isenabled())
gc.disable()
print(gc.isenabled())
gc.enable()
print(gc.isenabled())
```

```
True
False
True
```

10.4 Destructor

Python automatically deletes an object that is no longer in use. This automatic destroying of an object is known as garbage collection. Python periodically performs the garbage collection to free the blocks of memory that are no more required whereas the class can work with the special method _del_(), known as destructor, which gets invoked when the instance is almost destroyed.

Program: Write a program showing the cleanup activities of destructor.

```
import time
class A:
    def __init__(self):
        print("Constructor")
    def __del__(self):
        print("Destructor")
a1=A()
a2=a1
a3=a2
del a1
time.sleep(10)
print("object not destroyed a1")
del a2
time.sleep(10)
print("Object not destroyed a2")
print("Delete last reference")
del a3
```

```
Constructor
object not destroyed a1
Object not destroyed a2
Delete last reference
Destructor
```

If there are no reference variables associated with an object then it is valid for the garbage collector.

Example:

```
import time
class A:
    def __init__(self):
        print("Object Initialization")
    def __del__(self):
        print("Cleanning Activity")
a1=A()
a1=None
time.sleep(10)
print("End")
```

```
Object Initialization
Cleanning Activity
End
```

Example:

```
import time
class A:
    def __init__(self):
        print("Constructor")
    def __del__(self):
        print("Destructor")
l=[A(),A()]
del l
time.sleep(10)
print("End")
```

```
Constructor
Constructor
Destructor
Destructor
End
```

10.5 Finding the Number of References of an Object

For finding the number of references of an object, the module sys comprises the getrefcount() function.

Syntax: sys.getrefcount(objectreference)

Program: Write a program for finding the number of references of the object.

```
import sys
class A:
    pass
a1=A()
a2=a1
a3=a1
print(sys.getrefcount(a1))
```

4

10.6 Encapsulation

Encapsulation is used to combine the functions and data into a single entity. It is implemented using classes. The member function plays a vital role for accessing the data members of the class. The data is accessed using the functions of the class.

Example:

```
class Parent:
    def __init__(self):
        # Protected member
        self._a = 2
class Child(Parent):
    def __init__(self):
        # Calling constructor of Base class
        Parent.__init__(self)
        print("Calling protected member of parent class: ")
        print(self._a)
c = Child()
p = Parent()
# Calling protected member Outside class result in AttributeError
print(p.a)

Calling protected member of parent class:
2

AttributeError: 'Parent' object has no attribute 'a'
```

10.7 Inheritance

Inheritance is the method of generating a new class from an already defined class. The synonyms of an old class are the base class, super class, or the parent class whereas the synonyms of a new class are the sub class, derived class, or the child class. It permits the sub-classes to inherit entire variables and the methods to their associated parent class.

10.7.1 By Composition (HAS-A Relationship)

Composition is to access the members of one class in the other class by the class name or by creating its object.

Example:

```
class P:
    x=15
    def __init__(self):
        self.y=25
    def method1(self):
        print('P class data')
class Q:
    def __init__(self):
        self.p=P()
    def method2(self):
        print(self.p.x)
        print(self.p.y)
        self.p.method1()
q=Q()
q.method2()

15
25
P class data
```

Example:

```
class P:
    def __init__(self,name,eye,color):
        self.name=name
        self.eye=eye
        self.color=color
    def m1(self):
        print("Name:{} Eye:{} and Color:{}".format(self.name
                                ,self.eye,self.color))
class C:
    def __init__(self,cname,ear,head):
        self.cname=cname
        self.ear=ear
        self.head=head
    def m2(self):
        print("Name:",self.cname)
        print("Ear:",self.ear)
        print("P class Information:")
        self.head.m1()
p=P("Golu","Blue","Fair")
c=C("Shashwat",2,p)
c.m2()
```

```
Name: Shashwat
Ear: 2
P class Information:
Name:Golu Eye:Blue and Color:Fair
```

Example:

```
class AB:
    p=10
    def __init__(self):
        self.q=22
    def method1(self):
        print("method-1")
class CD:
    r=30
    def __init__(self):
        self.s=40
    def method2(self):
        print("method-2")
    def method3(self):
        ab=AB()
        print(ab.p)
        print(ab.q)
        ab.method1()
        print(CD.r)
        print(self.s)
        self.method2()
        print("method-3")
cd=CD()
cd.method3()
```

```
10
22
method-1
30
40
method-2
method-3
```

10.7.2 By Inheritance (IS-A Relationship)

The prime benefit of inheritance is the code reusability or extending the existing function-ality with the new functionality. There is no need to re-write the methods, variables, and constructors of the parent class to the sub-class if you want to extend the functionality of the parent class.

Syntax:　　　class child _ class(parent_class)

```python
class GOD:
    p=10
    def __init__(self):
        self.q=10
    def method1(self):
        print('Instance method-1')
    @classmethod
    def method2(cls):
        print('Class method-2')
    @staticmethod
    def method3():
        print('Static method-3')
class Moster(GOD):
    pass
m=Moster()
print(m.p)
print(m.q)
m.method1()
m.method2()
m.method3()
```

```
10
10
Instance method-1
Class method-2
Static method-3
```

Using inheritance, the members declared in the parent class will be present in the child class.

Example:

```python
class Side:
    def method1(self):
        print("Parent method-1")
class Guide(Side):
    def method2(self):
        print("Child method-2")
g=Guide()
g.method1()
g.method2()
```

```
Parent method-1
Child method-2
```

Example:

```
class Side:
    p=3
    def __init__(self):
        self.q=6
class Guide(Side):
    r=30
    def __init__(self):
        super().__init__()
        self.s=9
g=Guide()
print(g.p,g.q,g.r,g.s)
```

```
3 6 30 9
```

The methods present in the parent class will be present in the child class automatically; thus, when you refer to the child class, you can call both the parent and child class methods.

Example:

```
class Computer:
    def __init__(self,name,gen):
        self.name=name
        self.gen=gen
    def m1(self):
        print('Method m1')
class Mobile(Computer):
    def __init__(self,name,gen,ver,price):
        super().__init__(name,gen)
        self.ver=ver
        self.price=price
    def m2(self):
        print("Method m2")
    def m3(self):
        print("Name:",self.name)
        print("Generation:",self.gen)
        print("Version:",self.ver)
        print("Price:",self.price)
m=Mobile('Apple',5, 12.0, 100000)
m.m1()
m.m2()
m.m3()
```

```
Method m1
Method m2
Name: Apple
Generation: 5
Version: 12.0
Price: 100000
```

IS-A vs HAS-A Relationship

The IS-A relationship allows you to extend the functionality by adding new functionalities whereas the HAS-A relationship doesn't allow you to extend the functionality with the new functionalities; it allows us to use only the existing functionality.

Example: Here in Figure 10.1, the employee class is extending the functionality of the person class whereas the employee class is just using the functionality of company without extending it.

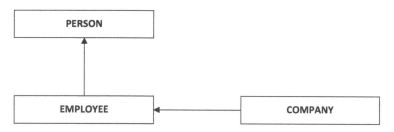

FIGURE 10.1
IS-A vs HAS-A relationship

Example:

```
class Man:
    def __init__(self,n,h,c):
        self.n=n
        self.h=h
        self.c=c
    def m1(self):
        print("Name:{} Height:{} Color:{}".format(self.n
                                       ,self.h, self.c))
class Women:
    def __init__(self,n,s):
        self.n=n
        self.s=s
    def m2(self):
        print('Method-2')
class Kids(Women):
    def __init__(self,n,s,g,a):
        super().__init__(n,s)
        self.g=g
        self.a=a
    def m3(self):
        print("Method-3")
    def m4(self):
        print("Name:",self.n)
        print("Status:",self.s)
        print("Gender:",self.g)

        #print("Age:",self.a)
        print("Man class Information:")
        self.a.m1()
m=Man("Ram","6ft","Dark")
k=Kids("Rani","Married","Female",m)
k.m2()
k.m3()
k.m4()
```

```
Method-2
Method-3
Name: Rani
Status: Married
Gender: Female
Man class Information:
Name:Ram Height:6ft Color:Dark
```

10.8 Aggregation vs Composition

Composition is the process of deploying the complex objects as the complex objects are objects that are deployed from the smaller or the simpler objects. In object-oriented programming languages, object composition is preferred for the objects that possess the HAS-A relationship with each other. Composition also says that without the existence of a container object, the contained object existence is not possible. Thus, the container and the contained object maintain a strong association relation among them. For example, consider the parent as the container and the child as the contained object. As the child can't exist without the parent, they are strongly associated with each other. In composition, complex classes have data members belonging to other simpler classes. If the associations among two objects are weak, and none of the objects has exclusive ownership of other, then it is not composition (Soh et al. 2009). It is rather called aggregation. Thus, the container and the contained object maintain a weak association relation among them. For example, consider the company as the container and the employee as the contained object. If the employee exists without company, they are weakly associated with each other. Composition always shows the relationship among objects and their instance variables whereas Aggregation always shows the relationship among objects and static variables.

Example:

```python
class DOC:
    x='Python'
    def __init__(self,n):
        self.n=n
        print(DOC.x)
d=DOC('Book')
print(d.n)
```

```
Python
Book
```

Note: The child class constructor gets processed, after creation of the child class object whereas the parent class constructor will be processed if the child class doesn't comprise the constructor, but the parent object will not be constructed.

Example:

```python
class COM:
    def __init__(self):
        print(id(self))
class DOM(COM):
    pass
d=DOM()
print(id(d))
```

```
2115164668688
2115164668688
```

Example:

```python
class Faculty:
    def __init__(self,name,deptt):
        self.name=name
        self.deptt=deptt
class Student(Faculty):
    def __init__(self,name,deptt,sub):
        super().__init__(name,deptt)
        self.sub=sub
    def __str__(self):
        return 'Name: {}\nDepartment: {}\nSubject: {}'
        .format(self.name,self.deptt,self.sub)
s=Student('Samarth',"CSE","Python")
print(s)
```

```
Name: Samarth
Department: CSE
Subject: Python
```

10.9 Inheritance

Inheritance is the procedure of deriving the new class from an existing class. Its types are mentioned below:

10.9.1 Single Inheritance

According to single inheritance, the properties of one class can be inherited to other class. In Figure 10.2, the properties of class P can be inherited by class S.

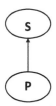

FIGURE 10.2
Single inheritance

Example:

```python
class Parent:
    def method1(self):
        print("Parent Method")
class Child(Parent):
    def method2(self):
        print("Child Method")
c=Child()
c.method1()
c.method2()
```

```
Parent Method
Child Method
```

10.9.2 Multilevel Inheritance

Inheritance that involves more than a single parent class but at different levels is known as multilevel inheritance. In Figure 10.3, the properties of class P can be inherited by class S and the properties of class S are inherited by class V.

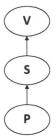

FIGURE 10.3
Multilevel inheritance

Example:

```python
class Parent:
    def method1(self):
        print("Parent Method")
class Child(Parent):
    def method2(self):
        print("Child Method")
class SubChild(Child):
    def method3(self):
        print("Sub Child Method")
c=SubChild()
c.method1()
c.method2()
c.method3()
```

```
Parent Method
Child Method
Sub Child Method
```

10.9.3 Hierarchical Inheritance

According to hierarchical inheritance, the properties of one class can be inherited by multiple classes of same level. In Figure 10.4, the properties of class P can be inherited by class S and class V.

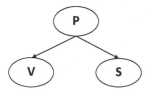

FIGURE 10.4
Hierarchical inheritance

Example:

```
class Parent:
    def method1(self):
        print("Parent Method")
class Child1(Parent):
    def method2(self):
        print("Child-1 Method")
class Child2(Parent):
    def method3(self):
        print("Child-2 Method")
c1=Child1()
c1.method1()
c1.method2()
c2=Child2()
c2.method1()
c2.method3()
```

```
Parent Method
Child-1 Method
Parent Method
Child-2 Method
```

10.9.4 Multiple Inheritances

According to multiple inheritances, it is possible to inherit from more than a single parent class. In this case all the methods and the attributes of both the parent class will be present in the child class after inheritance. In Figure 10.5, the properties of class S and class V are being inherited by class P.

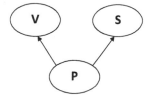

FIGURE 10.5
Multiple inheritance

Example:

```
class A:
    def method1(self):
        print("A class")
class B:
    def method2(self):
        print("B class")
class C(A,B):
    def method3(self):
        print("C class")
c=C()
c.method1()
c.method2()
c.method3()
```

```
A class
B class
C class
```

If the same method is inherited from both the parent classes, then Python will take care of the sequence of the parent classes while declaring the child class.

class P(S1, S2): S1 method is considered
class P(S2, S1): S2 method is considered

Example:

```
class A:
    def method1(self):
        print("A class")
class B:
    def method2(self):
        print("B class")
class C(A,B):
    def method3(self):
        print("C class")
c=C()
c.method1()
c.method2()
```

```
A class
B class
```

10.9.5 Hybrid Inheritance

Combining the single inheritance, multilevel inheritance, multiple inheritance, and hierarchical inheritance is called as hybrid inheritance. Figure 10.6 is an example of hybrid inheritance.

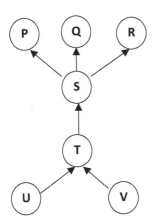

FIGURE 10.6
Hybrid inheritance

10.9.6 Cyclic Inheritance

According to cyclic inheritance, the properties of one class can be inherited in cyclic way to another class as shown in Figure 10.7.

FIGURE 10.7
Cyclic inheritance

Example:

```
class A(B):
    pass
class B(A):
    pass
NameError: name 'B' is not defined
```

10.10 Method Resolution Order (MRO)

In multilevel inheritance, any specific attribute is searched first in the main class. If it is not present there, then the first derived class is searched, if it is not found even there then the base class is searched. If the attribute is still not found, then finally the object class is checked. The order is known as the linearization of the derived class. Correspondingly, the set of protocols will be used to finding the linearization order as a Method Resolution Order (MRO). It guarantees that the classes come prior to their parent class. However, in multiple inheritance, MRO is similar to tuples of the base classes. The MRO of class can be checked either by using the _mro_ attribute or the mro() method.

Example:

```
class P:pass
class Q(P):pass
class R(P):pass
class S(Q,R):pass
print(P.mro())
print(Q.mro())
print(R.mro())
print(S.mro())

[<class '__main__.P'>, <class 'object'>]
[<class '__main__.Q'>, <class '__main__.P'>, <class 'object'>]
[<class '__main__.R'>, <class '__main__.P'>, <class 'object'>]
[<class '__main__.S'>, <class '__main__.Q'>, <class '__main__.R'>,
```

Example:

```
class A:pass
class B:pass
class C:pass
class X(A,B):pass
class Y(B,C):pass
class P(X,Y,C):pass
print(A.mro())
print(X.mro())
print(Y.mro())
print(P.mro())

[<class '__main__.A'>, <class 'object'>]
[<class '__main__.X'>, <class '__main__.A'>
[<class '__main__.Y'>, <class '__main__.B'>
[<class '__main__.P'>, <class '__main__.X'>
```

Example:

```
class First:
    def fun(self):
        print('First')
class Second:
    def fun(self):
        print('Second')
class Thrid:
    def fun(self):
        print('Thrid')
class Four(First,Second):
    def fun(self):
        print('Four')
class Five(Second,Thrid):
    def fun(self):
        print('Five')
class Six(Four,Five,Thrid):
    def fun(self):
        print('Six')
s=Six()
s.fun()
```

```
Six
```

In the example given above, the class Six fun() method will be observed. If the class Six does not contain the fun() method, then as per MRO, the class Four method will be observed. If class Four does not contain the fun() method, then the class First method will be observed and this process continues. The process of method resolution is: SixFourFirstFiveTwoThree

10.11 Super() Method

The super() method is used when the sub-class is needed to direct its immediate super-class. It is also used for calling the constructor, the _init_ method of super class from the child class.

Example:

```
class Human:
    def __init__(self,name,gender):
        self.name=name
        self.gender=gender
    def method(self):
        print('Name:',self.name)
        print('Gender:',self.gender)
class Author(Human):
    def __init__(self,name,gender,title,price):
        super().__init__(name,gender)
        self.title=title
        self.price=price
    def method(self):
        super().method()
        print('Title:',self.title)
        print('Price:',self.price)
a1=Author('Shashwat','Male','Python Programming',8000)
a1.method()
```

```
Name: Shashwat
Gender: Male
Title: Python Programming
Price: 8000
```

Example:

```
class Parent:
    var1=18
    def __init__(self):
        self.b=10
    def method1(self):
        print('Instance method-1')
    @classmethod
    def method2(cls):
        print('Class method-2')
    @staticmethod
    def method3():
        print('Static method-3')
class Child(Parent):
    var1=50
    def __init__(self):
        self.var2=100
        super().__init__()
        print(super().var1)
        super().method1()
        super().method2()
        super().method3()
c=Child()
```

```
18
Instance method-1
Class method-2
Static method-3
```

10.11.1 Calling Method of a Specific Super Class

It can be called as follows:

super(D, self).p1():   p1() method will be called of the super class of D.
A.p1(self):   A.p1(self):Class A will be called p1() method

Example:

```
class One:
    def method(self):
        print('ONE')
class Two(One):
    def method(self):
        print('TWO')
class Three(Two):
    def method(self):
        print('THREE')
class Four(Three):
    def method(self):
        print('FOUR')
class Five(Four):
    def method(self):
        One.method(self)
f=Five()
f.method()
```

```
ONE
```

Important points to note the regarding super() method:

Case-1: Using the super method(), self is preferred for accessing the parent class instance variables whereas the parent class static variables can be accessed by super().

```
class One:
    var1=12
    def __init__(self):
        self.var2=28
class Two(One):
    def method(self):
        print(super().var1)
        print(self.var2)
        #print(super().var2)#invalid
t=Two()
t.method()
```

```
12
28
```

Case-2: The parent class instance, static, and class method can be accessed from the child class constructor and the instance method by super().

```
class Parent:
    def __init__(self):
        print('Constructor')
    def method1(self):
        print('Instance method-1')
    @classmethod
    def method2(cls):
        print('Class method-2')
    @staticmethod
    def method3():
        print('Static method-3')
class Child(Parent):
    def __init__(self):
        super().__init__()
        super().method1()
        super().method2()
        super().method3()
    def method1(self):
        super().__init__()
        super().method1()
        super().method2()
        super().method3()
c=Child()
c.method1()
```

```
Constructor
Instance method-1
Class method-2
Static method-3
Constructor
Instance method-1
Class method-2
Static method-3
```

Case-3: From the child class, class method you cannot directly access the parent class instance method and constructor by using super() whereas the parent class static and class methods can be accessed.

```python
class Parent:
    def __init__(self):
        print('Constructor')
    def method1(self):
        print('Instance method-1')
    @classmethod
    def method2(cls):
        print('Class method-2')
    @staticmethod
    def method3():
        print('Static method-3')
class Child(Parent):
    @classmethod
    def method1(cls):
        #super().__init__()
        #super().method1()
        super().method2()
        super().method3()
Child.method1()
```

```
Class method-2
Static method-3
```

Example: (calling parent class instance methods and constructors from the child method of child class)

```python
class Parent:
    def __init__(self):
        print('Constructor')
    def method1(self):
        print('Instance method-1')
class Child(Parent):
    @classmethod
    def method2(cls):
        super(Child,cls).__init__(cls)
        super(Child,cls).method1(cls)
Child.method2()
```

```
Constructor
Instance method-1
```

Case-4: Special way to use the super() method in the child class static method.

```
class Parent:
    def __init__(self):
        print('Constructor')
    def method1(self):
        print('Instance method-1')
    @classmethod
    def method2(cls):
        print('Class method-2')
    @staticmethod
    def method3():
        print('Static method-3')
class Child(Parent):
    @staticmethod
    def method1():
        super().method1()--->invalid
        super().method2()--->invalid
        super().method3()--->invalid
Child.method1()

SyntaxError: invalid syntax
```

Case-5: Calling the parent class from the child class static method using super().

```
class Parent:
    @staticmethod
    def method1():
        print('Static method')
class Child(Parent):
    @staticmethod
    def method2():
        super(Child,Child).method1()
Child.method2()

Static method
```

10.12 Polymorphism

Polymorphism comprises poly and morph; poly means many and morph means forms. It has the capability to use the operator or the function in different forms. It means that the single function or the operator works distinctly on the data available to them. When it is applied over a function or a method depending on the available parameters, a specific form of function can be selected for processing.

10.12.1 Duck Typing Philosophy

In Python, the type can't be specified explicitly. On the basis of the value provided during runtime, the type will be assumed by default. Thus, Python is regarded as a dynamic typing programming language.

```
def m1(obj):
    obj.m2()
```

The type of object cannot be decided at the start. During runtime, any type can be passed. During execution time, if it walks like a duck and talks like a duck, then it should be a duck. This is the principle followed by Python known as the Duck Typing Philosophy of Python.

```python
class Man:
    def method1(self):
        print('Man')
class Women:
    def method1(self):
        print('Women')
class Boy:
    def method1(self):
        print('Boy')
class Kids:
    def method1(self):
        print('Kids')
def fun(x):
    x.method1()
l=[Man(),Women(),Boy(),Kids()]
for x in l:
    fun(x)
```

```
Man
Women
Boy
Kids
```

Example: (if x doesn't contain method1() then, an AttributeError will be raised)

```python
class Man:
    def method1(self):
        print('Man')
class Women:
    def method2(self):
        print('Women')
def fun(x):
    x.method1()
m=Man()
fun(m)
w=Women()
fun(w)
```

```
Man
```

```
AttributeError: 'Women' object has no attribute 'method1'
```

The above error can be solved by hasattr() function:

> hasattr(obj, 'attribute_name')

where, attribute_name can be a method or a variable name

Example:

```
class Man:
    def method1(self):
        print('Man')
class Women:
    def method1(self):
        print('Women')
class Boy:
    def method2(self):
        print('Boy')
def fun(x):
    if hasattr(x,'method-1'):
        x.method1()
    elif hasattr(x,'method-2'):
        x.method2()
m=Man()
fun(m)
w=Women()
fun(w)
b=Boy()
fun(b)
```

10.12.2 Overloading

This operator can be used for various reasons. There are three types of overloading.

10.12.2.1 Operator Overloading

Operator overloading is an important attribute of object oriented programming. The procedure of assigning special meaning to the operator is the operator overloading. In it, the same operator can be used for a number of different reasons like + operator, which is also used for concatenation of two strings. Internally, the + operator is implemented by the _ add_() method, known as the magic method, whereas the * operator is implemented by the _mul_() method, also applicable for string repetition.

Example:

```
class Boy:
    def __init__(self,toys):
        self.toys=toys
b1=Boy(20)
b2=Boy(30)
print(b1+b2)

TypeError: unsupported operand type(s) for +: 'Boy' and 'Boy'
```

Program: Write a program to show the use of overload '+' operator for class object.

```
class Boy:
    def __init__(self,toys):
        self.toys=toys
    def __add__(self,var):
        return self.toys+var.toys
b1=Boy(40)
b2=Boy(80)
print(b1+b2)
```
120

A list of operators and their associated magic methods is shown in Table 10.1.

TABLE 10.1

List of operators and their methods

S. No.	Operator	Magic methods
1	-	object.__sub__(self, other)
2	+	object.__add__(self, other)
3	/	object.__div__(self, other)
4	*	object.__mul__(self, other)
5	%	object._mod_(self,other)
6	//	object.__floordiv__(self, other)
7	+=	object.__iadd__(self, other)
8	**	object._pow_(self,other)
9	*=	object._imul_(self,other)
10	-=	object.__isub__(self, other)
11	//=	object.__ifloordiv__(self,other).
12	/=	object.__idiv__(self,other)
14	**=	object.__ipow__(self,other).
14	%=	object._imod_(self,other)
15	<=	object.__le__(self,other).
16	<	object.__lt__(self,other)
17	>=	object.__ge__(self,other).
18	>	object.__gt__(self,other)
19	!=	object.__ne__(self,other).
20	==	object.__eq__(self,other)

Overloading > and <= operators for worker class objects:

```python
class Worker:
    def __init__(self,name,sal):
        self.name=name
        self.sal=sal
    def __gt__(self,var):
        return self.sal>var.sal
    def __le__(self,var):
        return self.sal<=var.sal
w1=Worker("A",10000)
w2=Worker("B",20000)
print(w1>w2)
print(w1<w2)
print(w1<=w2)
print(w1>=w2)
```

```
False
True
True
False
```

Example: (overloading multiplication operator)

```python
class A:
    def __init__(self,sal):
        self.sal=sal
    def __mul__(self,var):
        return self.sal*var.days
class B:
    def __init__(self,days):
        self.days=days
a=A(500)
b=B(27)
print('Total Amount:',a*b)
```

```
Total Amount: 13500
```

10.12.2.2 *Method Overloading*

Overloaded methods are those methods in which two methods share the same name, but their arguments are different. If the programmer declares various methods with the same name and different number of arguments, then the last method will be considered in Python programming (Hall 2010). Method overloading is not possible in Python.

Example:

```
class A:
    def method1(self):
        print('No arg')
    def method1(self,x):
        print('One arg')
    def method1(self,x,y):
        print('Two-arg')
a=A()
#a.method1()
#a.method1(4)
a.method1(3,5)
```

```
Two-arg
```

Handling of overloaded methods

If a method is required with variable number of arguments then it can be handled with the default or variable number of arguments.

Example (with default arguments):

```
class Arth:
    def sum(self,x=None,y=None,z=None):
        if x!=None and y!= None and z!= None:
            print('3 arg:',x+y+z)
        elif x!=None and y!= None:
            print('2 arg:',x+y)
        else:
            print('Default')
a=Arth()
a.sum(3,4)
a.sum(2,5,7)
a.sum(5)
```

```
2 arg: 7
3 arg: 14
Default
```

Example (with variable number of arguments):

```
class Arth:
    def sum(self,*x):
        s=0
        for i in x:
            s=s+i
        print('Sum:',s)
a=Arth()
a.sum(2,5)
a.sum(3,5,8)
a.sum(8)
a.sum()
```

```
Sum: 7
Sum: 16
Sum: 8
Sum: 0
```

10.12.2.3 Constructor Overloading

In Python, constructor overloading is not allowed. If the multiple constructors are declared, then the latest one overwrites all the previous constructor.

Example:

```
class Arth:
    def __init__(self):
        print('No-Arg Constructor')
    def __init__(self,x):
        print('One-Arg constructor')
    def __init__(self,x,y):
        print('Two-Arg constructor')
#a1=Arth()
#a1=Arth(4)
a1=Arth(3,5)

Two-Arg constructor
```

Constructor with default arguments:

```
class Arth:
    def __init__(self,x=None,y=None,z=None):
        print('Constructor')
a1=Arth()
a2=Arth(1)
a3=Arth(2,3)
a4=Arth(3,4,6)

Constructor
Constructor
Constructor
Constructor
```

Constructor with variable number of arguments:

```
class Arth:
    def __init__(self,*x):
        print('Constructor')
a1=Arth()
a2=Arth(2)
a3=Arth(3,4)
a4=Arth(3,4,5)

Constructor
Constructor
Constructor
Constructor
```

10.12.3 Overriding

10.12.3.1 Method Overriding

Method overriding is required when we want to define our own functionality in the child class. This is possible by defining a method in the child class that has the same name, same arguments, and similar return type as a method in the parent class. When this method is called, the method defined in the child class is invoked and executed instead of the one in the parent class.

Example:

```
class Upper:
    def m1(self):
        print('Upper m1')
    def m2(self):
        print('Upper m2')
class Lower(Upper):
    def m2(self):
        print('lower m2')
l=Lower()
l.m1()
l.m2()
```

```
Upper m1
lower m2
```

The parent class method can be called from the overriding method of child class by the super() method.

```
class Upper:
    def m1(self):
        print('Upper m1')
    def m2(self):
        print('Upper m2')
class Lower(Upper):
    def m2(self):
        super().m2()
        print('lower m2')
l=Lower()
l.m1()
l.m2()
```

```
Upper m1
Upper m2
lower m2
```

Example (for constructor overriding):

```
class Parent:
    def __init__(self):
        print('Parent Constructor')
class Child(Parent):
    def __init__(self):
        print('Child Constructor')
c=Child()
```

```
Child Constructor
```

Example (calling of parent class constructor by super() method):

```
class Parent:
    def __init__(self,name):
        self.name=name
class Child(Parent):
    def __init__(self,name,deptt):
        super().__init__(name)
        self.deptt=deptt
    def method(self):
        print('Name:',self.name)
        print('Department:',self.deptt)
c1=Child('Vijay','CSE')
c1.method()
c2=Child('Swati','IT')
c2.method()
```

```
Name: Vijay
Department: CSE
Name: Swati
Department: IT
```

10.13 Abstract Class

An abstract class is a class that is specifically defined to lay a foundation for other classes that exhibits a common behavior or similar characteristics. It is primarily used only as a base class for inheritance. Since an abstract class is an incomplete class, users are not allowed to create its object. To use such a class, programmers must derive it keeping in mind that they would only be either using or overriding the features specified in that class.

Therefore, we see that the abstract class just serves as a template for other classes by defining a list of methods that the classes must implement. It makes no sense to instantiate an abstract class because all the method definitions are empty and must be implemented in a subclass.

Case 1: Creation of the object for the Abst class as it is a concrete class and doesn't contain any abstract method.

```
from abc import *
class Abst:
    pass
a=Abst()
```

Case 2: Creation of the object after deriving from the ABC class as it doesn't possess any abstract method.

```
from abc import *
class Abst(ABC):
    pass
a=Abst()
```

Case 3:

```
from abc import *
class Abst(ABC):
    @abstractmethod
    def method1(self):
        pass
a=Abst()

TypeError: Can't instantiate abstract class

Abst with abstract methods method1
```

Case 4: Creation of the object when the class possesses the abstract method.

```
from abc import *
class Abst:
    @abstractmethod
    def method1(self):
        pass
a=Abst()
```

Case 5:

```
from abc import *
class Abst:
    @abstractmethod
    def method1(self):
        print('Abstract')
a=Abst()
a.method1()

Abstract
```

In the child class, the parent class abstract method can be implemented or the child class can't be instantiated.

Case 1: Itis valid as no child class object is being created.

```
from abc import *
class Shash(ABC):
    @abstractmethod
    def m1(self):
        pass
class Viz(Shash): pass
```

Case 2:

```
from abc import *
class Shash(ABC):
    @abstractmethod
    def m1(self):
        pass
class Viz(Shash): pass
v=Viz()

TypeError: Can't instantiate abstract class Viz with abstract methods m1
```

If the abstract class is being extended and doesn't override its abstract method, then the child class is also abstract and then the instantiation is not allowed.

```python
from abc import *
class A(ABC):
    @abstractmethod
    def m1(self):
        pass
class B(A):
    def m1(self):
        return "B class m1"
class C(A):
    def m1(self):
        return "C class m1"
b=B()
print(b.m1())
c=C()
print(c.m1())

B class m1
C class m1
```

10.14 Abstract Method

Abstract methods are used when the programmer is not aware about the implementation but still declares the method or the abstract method contains only the declaration part and no implementation part. In Python, the abstract method is declared using the @ abstractmethod decorator as follows:

```
@abstractmethod
def pqr (self):
        pass
```

The @abstractmethod decorator is available in the abc module. Thus, the abc module has to be imported or an error message will be raised. The abc stands for 'Abstract Base Class Module'. The child classes are responsible for providing the implementation for the parent class abstract methods.

Example:

```python
class Abst:
    @abstractmethod
    def p1(self):
        pass

NameError: name 'abstractmethod' is not defined
```

Example:

```python
from abc import *
class Abst:
    @abstractmethod
    def p1(self):
        pass
```

10.15 Interface

The abstract class is an interface definition. In inheritance, we say that a class implements an interface if it inherits from the class which specifies that interface. In Python, we use the NotImplementedError to restrict the instantiation of a class. Any class that has the NotImplementedError inside method definitions cannot be instantiated.

Program: Write a program to show the concept of interface.

```
from abc import *
class DBInterface(ABC):
    @abstractmethod
    def connect(self):pass
    @abstractmethod
    def disconnect(self):pass
class DB2(DBInterface):
    def connect(self):
        print('Connecting to DB2')
    def disconnect(self):
        print('Disconnecting to DB2')
class Mysql(DBInterface):
    def connect(self):
        print('Connecting to Mysql')
    def disconnect(self):
        print('Disconnecting to Mysql')
dbn=input('Enter Database Name:')
cn=globals()[dbn]
var=cn()
var.connect()
var.disconnect()
```

```
Enter Database Name:Mysql
Connecting to Mysql
Disconnecting to Mysql
```

10.15.1 Concrete Class vs Abstract Class vs Interface

Interface is preferred when there is no information about implementation and only the requirement specification is known. The abstract class is preferred when you are talking about implementation but not completely. The concrete class is preferred when you are talking about implementation completely.

Program: Write a program to show the difference between the concrete class, abstract, and Interface.

```python
from abc import *
class CollegeAutomation(ABC):
    @abstractmethod
    def m1(self): pass
    @abstractmethod
    def m2(self): pass
    @abstractmethod
    def m3(self): pass
class AbsCls(CollegeAutomation):
    def m1(self):
        print('m1 method implementation')
    def m2(self):
        print('m2 method implementation')
class ConcreteCls(AbsCls):
    def m3(self):
        print('m3 method implemnentation')
c=ConcreteCls()
c.m1()
c.m2()
c.m3()
```

```
m1 method implementation
m2 method implementation
m3 method implemnentation
```

10.15.2 Public, Protected, and Private Attributes

The public attributes are those that can be accessed from within the class or from outside the class. By default, every attribute is public. The protected attributes are those that can be accessed both from outside and inside the class, only in child classes. An attribute can be specified by prefixing with the _ symbol.

Syntax: _varname = value

The private attributes are those that can be accessed only within the class. A variable can be declared as private explicitly by prefixing with two underscore symbols.

Syntax: _Varianbleme = value

Program: Show the use of public, private, and protected attributes.

```
class Boss:
    p=100
    _q=200
    __r=300
    def method1(self):
        print(Boss.p)
        print(Boss._q)
        print(Boss.__r)
b=Boss()
b.method1()
print(Boss.p)
print(Boss._q)
print(Boss.__r)

100
200
300
100
200

AttributeError: type object 'Boss' has no attribute '__r'
```

10.16 __str__() Method

Internally, the __str__() method will be called when you will be printing any object reference, which will return the string as follows:

<_main_.classname object at 0x022144B0>

For returning meaningful string representation, you have to override __str__() method.

Program: Write a program to show the use of the _str_() method.

```
class Gama:
    def __init__(self,name):
        self.name=name
    def __str__(self):
        return 'Name:{}'.format(self.name)
g1=Gama('vijay')
g2=Gama('parth')
print(g1)
print(g2)

Name:vijay
Name:parth
```

Example:

```
import datetime
t=datetime.datetime.now()
s=str(t)
print(s)
d=eval(s)

SyntaxError: leading zeros in decimal integer literals are not permitted
```

Program: Write a program to convert the date time object to string and the string object to date time using the repr() function.

```
import datetime
t=datetime.datetime.now()
p=repr(t)
print(p)
e=eval(p)
print(e)
```

```
datetime.datetime(2021, 1, 22, 1, 41, 23, 640670)
2021-01-22 01:41:23.640670
```

10.17 Conclusion

In this chapter, you have learned the advanced concepts of object oriented programming in Python. Generally, most of the modern programming languages follow the OOP principle, so the knowledge you gained in this chapter will be applicable throughout the programming arena.

Review Questions

1. How does inheritance allows users to reuse code?
2. Give the syntax of multiple inheritance.
3. What are abstract classes?
4. What do you understand by linearization?

Programming Assignments

PA 1: Write a program that has classes such as Student, Course and Department. Enroll a student in a course of a particular department.

PA2: Write a program that has a class Person. Inherit a class Faculty from Person that also has a class Publications.

PA 3: Write a program that extends the class Result so that the final result of the Student is evaluated based on the marks obtained in tests, activities, and sports.

PA 4: Define a class Employee. Display the personal and salary details of five employees using single inheritance.

References

Alchin, Marty. 2010. *Pro Python*. Apress.

Hall Tim, and J. P. Stacey. 2010. *Python 3 for absolute beginners*. Apress.

Soh Leen, Ashok Samal, Stephen Scott, Stephen Ramsay, Etsuko Moriyama, George Meyer, Brian Moore, William G. Thomas, and Duane F. Shell. 2009. Renaissance computing: an initiative for promoting student participation in computing. *ACM Technical Symposium on Computer Science Education*: 59–63.

Yeshwanthraj. 2019. Lecture notes on data structure for Lovely Professional University students. www.coursehero.com/file/50175645/A1802200285-23509-31-2019-listpdf/ (accessed December 4, 2021).

11

Exception Handling

11.1 Introduction

In this chapter, you will gain in-depth knowledge about exceptions, error, default exception handling, user or customized exception handling, and assertions in Python. The fine differences between error and exception have been explained with the help of examples. Errors are the problems in a program that will cause the program to stop the execution whereas exceptions are raised where some internal events occur that changes the normal flow of the program. Python provides a way to handle exceptions so that the code can be executed without any disturbance. Exception handling is a concept used in Python to handle the exceptions that occur during the execution of a program. If the exceptions are not handled, the interpreter will not be able to execute the code that exists after the exception. The main advantage of exception handlers is that they help to solve the problems that occur when resource allocation needs to reversed in a simple way.

11.2 Types of Error

In any programming language, we get errors. Basically, there are two kinds of errors, i.e. syntax errors and runtime errors (also known as exceptions). Sometimes, the program behaves in an abnormal fashion or in an unexpected way due to an error.

a) Syntax error: Syntax errors arise when a rule of the programming language is violated and it is the most basic error that generally occurs.

Example:

```
var = 12
if var == 12
print("MNNIT")

SyntaxError: invalid syntax
```

DOI: 10.1201/9781003185505-11

Example:

```
print "Python"

SyntaxError: Missing parentheses in call to 'print'
```

b) Runtime error: The other type of error is known as runtime error. Sometimes, the program gets executed but produces an incorrect result. This may occur due to incorrect logic in the designed algorithm.

Example:

```
#print(5/0) #division by zero
print(5/"one")

TypeError: unsupported operand type(s) for /: 'int' and 'str'
```

Example:

```
var = int(input("Enter Input:"))
print(var)

Enter Input:4
4

Enter Input:nine

ValueError: invalid literal for int() with base 10: 'nine'
```

11.3 Exception

A statement may cause an error during execution when it is not syntactically incorrect. Such errors that occur at runtime are known as exceptions. An event may occur while running the program and may even disturb the regular flow of the program, known as an exception. Whenever a program meets a condition that it cannot deal with, then an exception arises. Thus, in Python one can regard exception as an error (Chun 2001). Whenever a program declares an exception, it must be handled, or the program will be immediately terminated.

Syntax:

```
try:
        risky code
except:
        handling the encountered exception
```

11.4 Default Exception Handling

In Python, all the exceptions are treated as an object. Different classes are available, for all the exception types. When an exception is produced, PVM will manage the associated exception objects and then verify the code. If handling the code is not convenient then the Python interpreter discontinues the execution of the program in an abnormal way and displays the associated details of the exception on the console and the remaining program will not be processed further.

Example:

```
print("Hello")
print(10/0)
print("Hi")
```

```
Hello

ZeroDivisionError: division by zero
```

11.5 Customized Exception Handling: Using Try-Except

We can handle exceptions in our program by using a try block and except block. In the try block, the statements are kept that is critical in nature whereas in the except block, the code that will handle the exception is kept. It is also known as user defined exception handling.

try:
    statement
except ExceptionName:
    statement

Example: (without try-except)

```
print("First-1")
print(10/0)
print("Second-2")
```

```
First-1

ZeroDivisionError: division by zero
```

Example: (with try-except)

```
print("First-1")
try:
    print(10/0)
except ZeroDivisionError:
    print(10/2)
print("Second-2")

First-1
5.0
Second-2
```

Control flow of try-except:

Syntax: try:
                      statement-a
                      statement-b
                      statement-c
               except:
                      statement-d
                      statement-e

Case 1: When there is no exception raised then statement a, b, c, e will terminate normally.
Case 2: When an exception is raised in statement-b and its associated except block matches, then statement a, d, e will terminate normally.
Case 3: When an exception is raised at statement-b and its associated except block doesn't match, then statement-a will terminate normally.
Case 4: When an exception is raised at statement-d and statement-e then it will terminate abnormally.

Example: (To print exception information)

```
try:
    print(10/0)
except ZeroDivisionError as zde:
    print("exception occurs:",zde)

exception occurs: division by zero
```

11.6 Multiple Except Blocks

In Python, a number of except blocks can be taken for one try block. The block that matches with the exception generated will be executed. A try block can be associated with a number of except blocks for specifying the handlers for various exceptions (Lutz 2001). However, only a single handler will get executed for various exceptions.

Syntax: (one try block and a number of except blocks)

try:
        operations
        …………
except Exception1:
        If Exception1 occurs, this block will process
except. Exception2:
        If Exception2 occurs, this block will process
        …………………….
else:
        If no execution, this block will process …

If the try block is available with various except blocks, then on the basis of raised exception its associated except block will get process.

Example:

```
try:
    var1=int(input("Enter Var-1: "))
    var2=int(input("Enter Var-2: "))
    print(var1/var2)
except ZeroDivisionError :
    print("division by zero")
except ValueError:
    print("Only Integer")
```

```
Enter Var-1: 30
Enter Var-2: one
Only Integer
```

```
Enter Var-1: 30
Enter Var-2: 10
3.0
```

If the try block is available with various except blocks, then the order of the except block plays a vital role. The PVM will prefer the blocks from first to last until any of the exception conditions are met.

Example:

```
try:
    var1=int(input("Enter Var-1: "))
    var2=int(input("Enter Var-2: "))
    print(var1/var2)
except ArithmeticError:
    print("ArithmeticError")
except ZeroDivisionError:
    print("division by zero")
```

```
Enter Var-1: 40
Enter Var-2: 0
ArithmeticError
```

```
Enter Var-1: 9
Enter Var-2: 8
1.125
```

11.6.1 Multiple Exceptions in a Single Except Block

An except clause may name multiple exceptions as a parenthesized tuple, as shown in the program below. So, whatever exception is raised, out of the three exceptions specified, the same except block will be executed.

Example:

```
try:
    var1=int(input("Enter Var-1: "))
    var2=int(input("Enter Var-2: "))
    print(var1/var2)
except (ZeroDivisionError,ValueError) as msg:
    print("Invalid: ",msg)
```

```
Enter Var-1: 9
Enter Var-2: 8
1.125

Enter Var-1: 9
Enter Var-2: one
Invalid:  invalid literal for int() with base 10: 'one'

Enter Var-1: 9
Enter Var-2: 0
Invalid:  division by zero
```

11.6.2 Default Except Block

The default except block can be used for handling any variety of exception.

Syntax: except:
                statements

Example:

```
try:
    x=int(input("Enter First Number: "))
    y=int(input("Enter Second Number: "))
    print(x/y)
except ZeroDivisionError:
    print("Not divided by zero")
except:
    print("Default Except")
```

```
Enter First Number: 10
Enter Second Number: 0
Not divided by zero

Enter Var-1: 10
Enter Var-2: 9
1.111111111111112
```

11.7 The Finally Block

There is another block with the try block known as finally; it is an optional block. It is designed for performing the clean-up actions that are processed in all conditions. This block will always get processed before quitting the try block. This means that the statements written in the finally block get processed without considering the occurrence of exception.

Syntax:

try:
        risky code
except:
        handling code
finally:
        clean_up code

Case 1: When no exception occurs

```
try:
    print("try")
except:
    print("except")
finally:
    print("finally")

try
finally
```

Case 2: When an exception occurred but is handled

```
try:
    print("try")
    print(10/0)
except ZeroDivisionError:
    print("except")
finally:
    print("finally")

try
except
finally
```

Case 3: When an exception occurred but is not handled

```
try:
    print("try")
    print(10/0)
except NameError:
    print("except")
finally:
    print("finally")

try
finally

ZeroDivisionError: division by zero
```

Note: os._exit(0) is the only single condition in which the finally block doesn't execute and the PVM itself shuts down.

11.7.1 Control Flow in Try-Except-Finally

try:

        statement-a

        statement-b

        statement-c

except:

        statement-d

finally:

        statement-e

        statement-f

Case 1: When no exception raised then a, b, c, e, f will terminate normally.

Case 2: When an exception occurs at statement-b and its associated except block is matched then, a, d, e, f will terminate normally.

Case 3: When an exception occurs at statement-b but its associated except block didn't match then, a, e will terminate abnormally.

Case 4: When an exception occurs at statement-d then, it will always terminate abnormally but before terminating the finally block will be processed.

Case 5: When an exception occurs at statement-e or statement-f then it will always be terminated abnormally.

11.8 Nested Try-Except-Finally Block

The nested try-except-finally block can be taken as follows:

try:

    _____

    try:

        _____

    except:

        _____

except:

    _____

The risky code needs to be taken in the outer try block and the more risky code needs to be taken in the inner try block. If in the inner try block, if an exception occurs then, the inner except block will manage the exception. If it is unable to manage the exception then the outer except block will manage the exception.

Example:

```
try:
    print("outer try block")
    try:
        print("Inner try block")
        print(10/0)
    except ZeroDivisionError:
        print("Inner except block")
    finally:
        print("Inner finally block")
except:
    print("outer except block")
finally:
    print("outer finally block")
```

```
outer try block
Inner try block
Inner except block
Inner finally block
outer finally block
```

11.8.1 Control Flow in Nested Try-Except-Finally Block

```
try
        declaration-a
        declaration-b
        declaration-c
    try:
        declaration-d
        declaration-e
        declaration-f
except P:
    declaration-g
finally:
        declaration-h
        declaration-i
except Q:
        declaration-j
finally:
        declaration-k
        declaration-l
```

Case 1: When there is no exception, then a, b, c, d, e, f, h, i, k, l will terminate normally.

Case 2: When an exception occurs at declaration-b and its associated except block gets matched then a, j, k, l will terminate normally.

Case 3: When exception occurs at declaration-b and its associated except block didn't match, then a, k will terminate abnormally.

Case 4: When exception occurs at declaration-e and the inner except block gets matched then a, b, c, d, g, h, i, k, l, will terminate normally.

Case 5: When an exception occurs at declaration-e and the inner except block didn't match but the outer except block gets matched then a, b, c, d, h, j, k, l will terminate normally.

Case 6: When an exception occurs at declaration-e and the inner and the outer except blocks didn't match then a, b, c, d, h, k will terminate abnormally.

Case 7: When an exception occurs at declaration-g and its associated except block gets matched then, a, b, c,......, h, j, k, l will terminate normally.

Case 8: When an exception occurs at declaration-g and its associated except block didn't match then, a, b, c, d, e, f, g, h, k will terminate abnormally.

Case 9: When an exception occurs at declaration-h and its associated except block gets matched then a, b, c, ..., j, k, l will terminate normally.

Case 10: When exception occurs at declaration-h and its associated except block didn't match then a, b, c, ..., k will terminate abnormally.

Case 11: When an exception occurs at declaration-i and its associated except block gets matched then a, b, c, ..., h, j, k, l will terminate normally.

Case 12: When an exception occurs at declaration-i and its associated except block didn't match then a, b, c, ... h, k will terminate abnormally.

Case 13: When exception occurs at declaration-j then it will terminate abnormally but prior to abnormal termination, the finally block will be processed.

Case 14: When an exception occurs at declaration-k or declaration-l then it will terminate abnormally.

11.8.2 Else Block with Try-Except-Finally

The else block can be used with the block of try-except-finally. It gets executed only when the try block is free from exception.

```
try:
        Risky Code
except:
        executes when exception occurs
else:
        executes when no exception occurs
finally:
        always executes
```

Example:

```
try:
    print("try")
except:
    print("except")
else:
    print("else")
finally:
    print("finally")

try
else
finally
```

Example:

```
try:
    print("try")
    print(10/0)
except:
    print("except")
else:
    print("else")
finally:
    print("finally")

try
except
finally
```

11.9 Types of Exception

In Python, the following exceptions may occur.

1. In-built or pre-defined exception: Sometimes, the PVM will automatically raise an exception on the occurrence of an event, which is known as an in-built exception.

2. Customized or programmatic or user-defined exception: Sometimes, when the user derives their own class using standard in-built exceptions, new exceptions get created. The user-defined block occurs in the try block whereas it is caught in the except block.

11.10 Raise User-Defined Exception

Python allows programmers to re-raise an exception. For example, an exception thrown from the try block can be handled as well as re-raised except a block using the keyword raise.

```
class VIZException(Exception):
    def __init__(self,arg):
        self.msg=arg
class VIMException(Exception):
    def __init__(self,arg):
        self.msg=arg
age=int(input("Enter Age:"))
if age>60:
    raise VIZException("Senior Citizens")
elif age<18:
    raise VIMException("Child/Young")
else:
    print("Default Exception")
```

```
Enter Age:34
Default Exception

Enter Age:14
VIMException: Child/Young

Enter Age:98
VIZException: Senior Citizens
```

11.11 Assertion

An assertion is a basic check that can be turned on or off when the program is being tested. You can think of assert as a raise-if statement. Using an assert statement, an expression is being tested, and if the result of the expression is False then an exception is raised. The assert statement is intended for debugging statements. It can be seen as an abbreviated notation for a conditional raise statement.

In Python, assertions are implemented using an assert keyword. Assertions are usually placed at the start of a function to check for valid input, and after a function call to check for a valid output. When Python encounters an assert statement, the expression associated with it is calculated and if the expression is False, an AssertionError is raised.

Syntax: assert expression[,arguments]

If the assertion fails, Python uses ArgumentExpression as the argument for the AssertionError. AssertionError exceptions can be caught and handled like any other exception using the try-except block. However, if the AssertionError is not handled by the program, the program will be terminated, and an error message will be displayed.

Program: Write a program to calculate the square of any number using an assert command.

```
def fun(a):
    return a**a
assert fun(5)==25,"Square of 5: 25"
assert fun(6)==36,"Square of 6: 36"
assert fun(7)==49,"Square of 7: 49"
print(fun(5))
print(fun(6))
print(fun(8))

AssertionError: Square of 5: 25

def fun(a):
    return a*a
assert fun(5)==25,"Square of 5: 25"
assert fun(6)==36,"Square of 6: 36"
assert fun(7)==49,"Square of 7: 49"
print(fun(5))
print(fun(6))
print(fun(7))

25
36
49
```

11.12 Conclusion

Two important attributes of Python programming are used to handle any unexpected errors and to add to the debugging capabilities; exception handling and assertions have been explained in this chapter with a sufficient number of examples.

Review Questions

1. What is the purpose of exception handling?
2. Differentiate between an error and an exception.
3. What happens when an exception is raised in a program?
4. How can you handle multiple exceptions in a program?
5. Explain how you can instantiate an exception.

Programming Assignments

PA 1: Write a program that opens a file and writes data to it. Handle exceptions that can be generated during the I/O operations.

PA 2: Write a program that deliberately raises a user defined SocketError with any number of arguments and derived from the runtime class.

PA 3: Write a program that prompts a user to enter a number. If the number is positive or zero print it, otherwise raise an exception.

PA 4: Write a program that randomly generates a number and raises a user defined exception if the number is smaller than 0.1

References

Chun, Wesley. 2001. *Core Python programming*. Vol. 1. Prentice Hall Professional.

Lutz, Mark. 2001. *Programming Python*. O'Reilly Media.

12

File Handling

12.1 Introduction

A group of records is known as a file whereas a collection of interrelated data items is known as a record. The data items can contain any information, say students, employees, customers, etc. or a group of letter, numbers, symbols, characters, known as a file. A file is basically used because real-life applications involve a large amount of data and in these conditions the console-oriented input/output operations face the following difficulties:

- First, it is very difficult and time-consuming to handle such a bulk of data using terminals.
- Second, while performing I/O using terminals, the data gets lost on the termination of the program or by switching off the system. Thus, it is essential to save the data using permanent storage.

12.1.1 Opening and Closing a File

Before reading from or writing to a file, you first open by using the open() function. It is used to create the file object, which will be used to invoke its associated methods.

Syntax: fileobj = open(file_name [, access_mode])

While opening a file, the mode can be specified as 'r' for read, 'w' for write or 'a' to append to the file. The text or the binary mode can also be specified while opening the file (Hunt 2019). By default, the file is opened in text mode. Table 12.1 shows the different allowed modes with their purposes.

TABLE 12.1

Modes allowed on a file

Mode	Purpose
r	It is the default mode. The file gets opened for reading purpose only and the cursor points at the starting of the file.
w	The file gets opened for writing. If the file exists of the mentioned name, it gets overwritten else another file will be created.
a	The existing file is opened for the appending purpose. If the file exists, the pointer points at the end of file as the file is in the append mode otherwise, a new file will be created for the writing purpose.
r+	The file gets open for reading and writing as well. The existing data is the file doesn't gets deleted. The cursor is kept at the starting of file.
w+	File gets open for both reading and writing as well. It will overwrite the file if the file exists. Otherwise, another file will be deployed.
a+	The file gets open for append and read purpose. If the file exists, the pointer is at the end of file. The file opens in append mode whereas if the file doesn't exist, a new file will be created for read and write purpose.
x	The file gets open for exclusive creation for writing the data. The operation fails if the file already exists.

12.1.2 Attributes of a File Object

Table 12.2 shows the different attributes of files and the information obtained from them.

TABLE 12.2

File object attributes

Attribute	Information Obtained
Name	File name will be returned.
Mode	The access mode with which the file has opened will be returned.
Closed	If the file is closed, it will return True. Else False will be returned.
readable()	It will return the Boolean value mentioning whether the file is readable or not.
writable()	It will return the Boolean value mentioning whether the file is writable or not.

Example:

```
f=open("abc.txt",'w')
print("File Name: ",f.name)
print("File Mode: ",f.mode)
print("File Readable: ",f.readable())
print("File Writable: ",f.writable())
print("File Closed : ",f.closed)
f.close()
print("File Closed : ",f.closed)

File Name:   abc.txt
File Mode:   w
File Readable:   False
File Writable:   True
File Closed :   False
File Closed :   True
```

12.1.3 Writing and Reading Data to Text Files

To write a file, open it in append mode 'a', write 'w' or exclusive-creation 'x' mode. The write() method is used for writing the string or the sequence of bytes. It will inform you of the number of characters written in file. It doesn't add the new line character ('\n') at the end of string (Bell 2017). There are two methods to write the character data to the text file, write() and writelines().

Syntax: fileobject.write(string)
Syntax: fileobject.writelines(string)

Example: (data will be overridden after running the program)

```
f=open("viz.txt",'w')
f.write("Vijay\n")
f.write("MNNIT\n")
f.write("Allahabad\n")
print("Write data successfully")
f.close()
```

```
Write data successfully
```

After successful execution, a viz.txt file opens, which contains the following output:

> Vijay
> MNNIT
> Allahabad

If you don't want the data to be overwritten, then open the file as: x = open("viz.txt", "a")

Example:

```
f=open("viz.txt",'w')
l=["Shashwat\n","Vimal\n","Vijay\n","Swati"]
f.writelines(l)
print("Write data successfully")
f.close()
```

```
Write data successfully
```

After successful execution, a viz.txt file opens, which contains the following output:

> Shashwat
> Vimal
> Vijay
> Swati

Table 12.3 shows the various read methods to read the character data from the text file:

TABLE 12.3
Read methods

Methods	Information Obtained
read()	For reading the total data of file.
read(n)	For reading the 'n' characters of file
readline()	For reading a single line.
readlines()	For reading all the lines into a list.

Program: Write some Python code that reads the complete data of a file.

```
f=open("viz.txt",'r')
d=f.read()
print(d)
f.close()
```

```
Shashwat
Vimal
Vijay
Swati
```

Program: Write a program that will read the first 12 characters from a file.

```
f=open("viz.txt",'r')
d=f.read(12)
print(d)
f.close()
```

```
Shashwat
Vim
```

Program: Write some Python code that reads data line by line.

```
f=open("viz.txt",'r')
l1=f.readline()
print(l1,end='')
l2=f.readline()
print(l2,end='')
l3=f.readline()
print(l3,end='')
f.close()
```

```
Shashwat
Vimal
Vijay
```

Program: Write some Python code that reads all lines in a list.

```
f=open("viz.txt",'r')
l=f.readlines()
for i in l:
    print(i,end='')
f.close()
```

```
Shashwat
Vimal
Vijay
Swati
```

12.2 With Statement

This is used while opening a file. It is preferred for grouping the different file operations inside a block. The advantage is you don't need to take care to close the file explicitly; the file is closed automatically, and also in case of an exception.

Example:

```
with open("viz.txt","w") as f:
    f.write("Vijay\n")
    f.write("MNNIT\n")
    f.write("Allahabad\n")
    print("File Closed: ",f.closed)
print("File Closed: ",f.closed)

File Closed:   False
File Closed:   True
```

12.3 The Seek() and Tell() Methods

12.3.1 Tell() Method

The tell() method is used for returning the present location of the cursor to the start of the file. Like a string index, the index of the first character in the file is also zero.

Example:

```
f=open("viz.txt","r")
print(f.tell())
print(f.read(2))
print(f.tell())
print(f.read(3))
print(f.tell())

0
Vi
2
jay
5
```

12.3.2 Seek() Method

The seek() method is used to set the file pointer to a specific position in a file.

Syntax: File_object.seek(offset, attribute)

where, offset represents the number of position to be moved from the current position of the pointer and attribute indicates the point of reference from where the bytes are to be moved from.

The values for attribute are

> 0: The position is related to the start of the file.
> 1: The position is related to the current position.
> 2: The position is related to the end of the file.

Example:

```
d="Dr. Kalam was President of India"
f=open("viz.txt","w")
f.write(d)
with open("viz.txt","r+") as f:
    t=f.read()
    print(t)
    print("Current Cursor Position: ",f.tell())
    print("After Change the seek() position:")
    f.seek(14)
    print("Current Cursor Position: ",f.tell())
    f.write("|a Good Human being|")
    f.seek(0)
    t=f.read()
    print(t)
```

```
Dr. Kalam was President of India
Current Cursor Position:    32
After Change the seek() position:
Current Cursor Position:    14
Dr. Kalam was |a Good Human being|
```

12.4 Testing the Existence of a File

The os library is preferred for retrieving information about the files stored on the system. The os module has the path as a sub-module comprising the function is File(), which is used to check the existence of a file.

Syntax: os.path.isfile(fname)

Program: Write some Python code to check the existence of a file; if it is present then display its content.

```python
import os,sys
fn=input("Enter File Name: ")
if os.path.isfile(fn):
    print("File Here:",fn)
    f=open(fn,"r")
else:
    print("File does not exist:",fn)
    sys.exit(0)
    print("File Data:")
d=f.read()
print(d)
```

```
Enter File Name: viz.txt
File Here: viz.txt
Shashwat
Vimal
Vijay
Swati

Enter File Name: abc.txt
File does not exist: abc.txt
```

Program: Write a program to display the number of characters, words, lines available in the provided file.

```python
import os,sys
fn=input("Enter File Name: ")
if os.path.isfile(fn):
    print("File exist:",fn)
    f=open(fn,"r")
else:
    print("File does not exist:",fn)
    sys.exit(0)
lc=wc=cc=0
for l in f:
    lc=lc+1
    cc=cc+len(l)
    w=l.split()
    wc=wc+len(w)
print("Number of Lines Count:",lc)
print("Number of Words Count:",wc)
print("Number of Characters Count:",cc)
```

```
Enter File Name: abc.txt
File exist: abc.txt
Number of Lines Count: 3
Number of Words Count: 15
Number of Characters Count: 80

Enter File Name: xyz.txt
File does not exist: xyz.txt
```

12.5 Handling Binary Data

Reading or writing binary data such as audio, image, video files, etc. is an essential task that needs to be done regularly on files.

Program: Write a program for reading an image and writing it into a new image file.

```
f1=open("viz.jpg","rb")
f2=open("newpic.jpg","wb")
bytes=f1.read()
f2.write(bytes)
print("New Image with the new name: newpic.jpg")

New Image with the new name: newpic.jpg
```

12.6 Handling CSV Files

Reading and writing the data with respect to comma separated values (CSV) files, is a basic requirement of programming. There is a csv module to manage these csv files.

12.6.1 Writing Data to a CSV File

In csv files, the blank lines will be used among the data if you do not prefer the new-line attribute (Zammetti 2013). In Python-3 a new-line attribute is needed to prevent the blank lines.

Example:

```
import csv
with open("emp.csv","w",newline='') as f:
    w=csv.writer(f) # returns csv writer object
    w.writerow(["ENO","ENAME","ESAL","EADDR"])
    n=int(input("Enter Number of Employees:"))
    for i in range(n):
        eno=input("Enter Employee No:")
        ename=input("Enter Employee Name:")
        esal=input("Enter Employee Salary:")
        eaddr=input("Enter Employee Address:")
        w.writerow([eno,ename,esal,eaddr])
print("Employees data written in to csv file successfully")
```

```
Enter Number of Employees:2
Enter Employee No:100
Enter Employee Name:vijay
Enter Employee Salary:100000
Enter Employee Address:jaunpur
Enter Employee No:200
Enter Employee Name:vimal
Enter Employee Salary:200000
Enter Employee Address:allahabad
Employees data written in to csv file successfully
```

12.6.2 Reading Data from a CSV File

```
import csv
f=open("emp.csv",'r')
r=csv.reader(f) #returns csv reader object
d=list(r)
#print(d)
for l in d:
    for w in l:
        print(w,"\t",end='')
    print()
```

```
ENO       ENAME    ESAL      EADDR
100       vijay    100000    jaunpur
200       vimal    200000    allahabad
```

12.7 Zipping and Unzipping Files

Files are zipped to improve the utilization of memory, reducing the transportation time and improving the performance. To perform the zip and the unzip operations, there is an inbuilt module named zip file and a class named ZipFile in Python.

12.7.1 To Create a Zip File

The zip() is an inbuilt function in Python. It takes items in sequence from a number of collections to make a list of tuples, where each tuple contains one item from each collection. The function is often used to group items from a list that has the same index.

f = ZipFile("files.zip", "w", "ZIP_DEFLATED")

After the creation of a ZipFile object, the list of items is added as follows:

f.write(file_name)

Example:

```
from zipfile import *
f=ZipFile("f.zip",'w',ZIP_DEFLATED)
f.write("f1.txt")
f.write("f2.txt")
f.write("f3.txt")
f.close()
print("f.zip file created successfully")
```

```
f.zip file created successfully
```

12.7.2 To Perform Unzip Operations

The ZipFile object is created as:

f = ZipFile("files.zip", "r", ZIP_STORED)

ZIP_STORED denotes the unzip operation. There is no need to specify its value. After creating the ZipFile object for the unzip operation, all the names of the files available in the corresponding zip file can be retrieved by namelist() method.

names = f.namelist()

Example:

```
from zipfile import *
f=ZipFile("ff.zip",'r',ZIP_STORED)
n=f.namelist()
for n1 in n:
    print( "File Name: ",n1)
    print("File data is:")
    f1=open(n1,'r')
    print(f1.read())
    print()
```

```
File Name:  f1.txt
File data is:
Vijay is a good programmer

File Name:  f2.txt
File data is:
Hi this is vimal

File Name:  f3.txt
File data is:
hi This is Swati
```

12.8 Directory

A directory is a collection of files where each file may be of a different format. Python has various methods in the OS module that help programmers to work with directories (Gardner 2009).

12.8.1 Operations on a Directory

1. Identification of the current working directory

```
import os
cwd = os.getcwd()
print("Current Working Directory:",cwd)
```

```
Current Working Directory: C:\Users\swati sharma
```

2. Creation of a sub-directory inside the current working directory

```
import os
os.mkdir("mysub")
print("mysub directory created in cwd")
```

```
mysub directory created in cwd
```

3. Creation of a sub-directory inside the mysub directory

```
import os
os.mkdir("mysub/mysub2")
print("mysub2 created inside mysub")
```

```
mysub2 created inside mysub
```

4. Creation of multiple directories

```
import os
os.makedirs("sub1/sub2/sub3")
print("sub1 and in that sub2 and in that sub3 directories created")
```

```
sub1 and in that sub2 and in that sub3 directories created
```

5. To remove a directory

```
import os
os.rmdir("mysub/mysub2")
print("mysub2 directory deleted")
```

```
mysub2 directory deleted
```

6. To remove multiple directories in the path

```
import os
os.removedirs("sub1/sub2/sub3")
print("All 3 directories sub1,sub2 and sub3 removed")
```

```
All 3 directories sub1,sub2 and sub3 removed
```

7. To re-name a directory

```
import os
os.rename("mysub","newdir")
print("mysub directory renamed to newdir")
```

```
mysub directory renamed to newdir
```

8. Identification of the contents of a directory

The OS module provide the listdir() for listing the data of the specific directory. It will not show the contents of the sub-directory.

```
import os
print(os.listdir("."))
```

```
['.anaconda', '.android', '.AndroidStudio4.0', '.conda', '.condarc', '.config',
python', '.jupyter', '.keras', '.labelImgSettings.pkl', '.matplotlib', '.node_r
'abc.log', 'abc.txt', 'abcd.txt', 'Advance OOPs.ipynb', 'AndroidStudioProjects'
b', 'Breast Cancer', 'Calculator', 'config.txt', 'Contacts', 'Control Flow.ipyn
y', 'database.db', 'Database.ipynb', 'deep.jpg', 'Desktop', 'Documents', 'done.
'Employee.txt', 'ExceptionHandling.ipynb', 'f.zip', 'f1.txt', 'f2.txt', 'f3.txt
'Function.ipynb', 'getCustomLogger', 'gift_dataset5.db', 'IntelGraphicsProfiles
l Settings', 'logging.conf', 'Logging.ipynb', 'miet.png', 'Mini Project.ipynb',
g.txt', 'NetHood', 'New Text Document.txt', 'newpic.jpg', 'new_notebook.ipynb',
G2', 'NTUSER.DAT{fd9a35da-49fe-11e9-aa2c-248a07783950}.TxR.0.regtrans-ms', 'NTU
```

9. Identification of contents of a directory and sub-directory.

The walk() function is used to identify the contents of directory and sub-directory

Syntax: os.walk(path,topdown = True, onerror = None, followlinks = False)

It returns an iterator object and its contents can be seen using the for loop.

- path: Directory Path
- If topdown = True, then it will be traversed from top-to-bottom.
- If onerror = None, then it returns which function has to be executed.
- If followlinks = True, then the directories will be visited as indicated by the symbolic links.

Program: Write a program to print the contents of the current working directory and its associated sub-directories.

```
import os
for dirpath,dirnames,filenames in os.walk('.'):
    print("Current Directory Path:",dirpath)
    print("Directories:",dirnames)
    print("Files:",filenames)
    print()
```

```
Current Directory Path:  .
Directories: ['.anaconda', '.android', '.AndroidStudio4.0', '.conda',
'.ipython', '.jupyter', '.keras', '.matplotlib', '.PyCharmCE2019.3', '
ation Data', 'Breast Cancer', 'Calculator', 'Contacts', 'Cookies', 'De
ction', 'Favorites', 'getCustomLogger', 'IntelGraphicsProfiles', 'kera
'NetHood', 'OneDrive', 'Pack', 'Pack1', 'Pack2', 'Parth', 'Pictures',
cent', 'Saved Games', 'Searches', 'SendTo', 'source', 'Start Menu', 'T
Files: ['.condarc', '.labelImgSettings.pkl', '.node_repl_history', 'ab
sic OOP.ipynb', 'config.txt', 'Control Flow.ipynb', 'Data Structure.ip
jpg', 'done.txt', 'emp.csv', 'emp.dat', 'Employee.txt', 'ExceptionHand
zip', 'File Handling.ipynb', 'Function.ipynb', 'gift_dataset5.db', 'In
t.png', 'Mini Project.ipynb', 'Modules.ipynb', 'mylog.txt', 'New Text
```

Example:

```
import os
os.system("dir *.py")
os.system("py abc.py")
```

1

12.9 To Get Information about a File

The OS module stat() function is used to identify the statistics of a file like its size, its last accessed time, its last modified time, and so on. Table 12.4 shows the different parameters of a file and the information obtained from them.

Syntax: stats = os.stat("abc.txt")

TABLE 12.4
File parameters

Parameters	Information Obtained
st_mode	Protection Bits.
st_dev	Device..
st_ino	Node Number.
st_uid	User Id of Owner.
st_nlink	Number of Hard Links..
st_size	Size of file in bytes.
st_gid	Group-Id of Owner.
st_atime	Time of Most Recent Access.
st_ctime	Time of Most Recent Meta Data Change.
st_mtime	Time of Most Recent Modification.

12.9.1 Displaying Statistics of a File

```
import os
stats=os.stat("viz.txt")
print(stats)
```

```
os.stat_result(st_mode=33206, st_ino=9570149208180121, st_dev=1019728671
1611309479, st_mtime=1611308838, st_ctime=1608411891)
```

12.9.2 To Print Specified Properties

```
import os
from datetime import *
s=os.stat("abc.txt")
print("File Size in Bytes:",s.st_size)
print("File Last Accessed Time:",datetime.fromtimestamp(s.st_atime))
print("File Last Modified Time:",datetime.fromtimestamp(s.st_mtime))
```

```
File Size in Bytes: 82
File Last Accessed Time: 2020-12-21 23:50:09.614851
File Last Modified Time: 2020-12-21 12:30:41.895920
```

12.10 Pickling and Unpickling of Objects

Pickling is the procedure of writing the state of an object to the file whereas unpicking is the procedure of reading the state of an object from the file. They are implemented using a Python module named pickle. It comprises a dump() function for pickling.

> pickle.dump(object,file)

It comprises a load() function for unpickling.

> obj = pickle.load(file)

12.10.1 Reading and Writing the State-of-Objects

```python
import pickle
class Employee:
    def __init__(self,eno,ename,esal,eaddr):
        self.eno=eno;
        self.ename=ename;
        self.esal=esal;
        self.eaddr=eaddr;
    def display(self):
        print(self.eno,"\t",self.ename,"\t",self.esal,"\t",self.eaddr)
with open("emp.dat","wb") as f:
    e=Employee(1000,"Swati",10000,"Delhi")
    pickle.dump(e,f)
    print("Pickling of Employee Object completed")
with open("emp.dat","rb") as f:
    obj=pickle.load(f)
    print("Employee data after unpickling")
    obj.display()
```

```
Pickling of Employee Object completed
Employee data after unpickling
1000     Swati    10000    Delhi
```

12.10.2 Writing Multiple Objects

emp.py

```python
class Employee:
    def __init__(self,eno,ename,esal,eaddr):
        self.eno=eno
        self.ename=ename
        self.esal=esal
        self.eaddr=eaddr
    def display(self):
        print(self.eno,"\t",self.ename,"\t",self.esal,"\t",self.eaddr)
```

pick.py:

```python
import emp,pickle
f=open("emp.dat","wb")
n=int(input("Enter The number of Employees:"))
for i in range(n):
    eno=int(input("Enter Employee Number:"))
    ename=input("Enter Employee Name:")
    esal=float(input("Enter Employee Salary:"))
    eaddr=input("Enter Employee Address:")
e=emp.Employee(eno,ename,esal,eaddr)
pickle.dump(e,f)
print("Employee Objects pickled successfully")
```

```
Enter The number of Employees:1
Enter Employee Number:1020
Enter Employee Name:Swati
Enter Employee Salary:50000
Enter Employee Address:Delhi
Employee Objects pickled successfully
```

unpick.py

```
import emp,pickle
f=open("emp.dat","rb")
print("Employee Details:")
while True:
    try:
        obj=pickle.load(f)
        obj.display()
    except EOFError:
            print("All employees Completed")
            break
f.close()
```

```
Employee Details:
All employees Completed
```

```
Enter The number of Employees:2
Enter Employee Number:001
Enter Employee Name:vijay
Enter Employee Salary:100000
Enter Employee Address:jaunpur
Enter Employee Number:002
Enter Employee Name:deepak
Enter Employee Salary:20000
Enter Employee Address:kashipur
Employee Objects pickled successfully
Enter The number of Employees:
```

12.11 Conclusion

In this chapter, you have studied how Python handles two of the most important components of any operating system: the files and directories. Fortunately, Python has in-built functions to create and manipulate flat and text files. We can conclude that in comparison to any other programming language, file handling is quite simple in Python.

Review Questions

1. Differentiate between an absolute and relative path.
2. Explain the utility of the open() function.
3. Give the significance of with keyword.
4. Explain the attributes of a file object.
5. Differentiate between a file and a folder.

Programming Assignments

PA 1: Write a program that reads a file and prints only those lines that have the word 'print'.

PA2: Write a program to compare two files.

PA 3: Write a program to count the number of records stored in a file employee.

PA 4: Write a program that exchanges the contents of two files.

References

Bell, Charles. 2017. How to program in MicroPython. In *MicroPython for the Internet of Things*. Apress, Berkeley, CA: 125–171.

Gardner, Scott. 2009. *The definitive guide to pylons*. Apress.

Hunt, John. 2019. Reading and writing files. In *Advanced guide to Python 3 programming*. Springer: 215–230.

Zammetti, Frank. 2013.*Learn Corona SDK game development*. Apress.

13

Multithreading

13.1 Introduction

Before learning multithreading, let us first understand multitasking in Python.

Multitasking is the execution of several tasks simultaneously. The types of multitasking are mentioned below:

a) Process-based multitasking: Execution of several jobs at a time where all the jobs are running on a single processor is process-based multitasking. For example, we may be typing our résumé, downloading a movie over the internet, and listening to songs from the same computer at the same time. This is an example of simultaneous and independent execution of tasks. It is most suitable at OS level.

b) Thread-based multitasking: Execution of various jobs at a time where all the jobs are the individual part of the same program is known as thread-based multitasking and all the individual parts are called as threads. It is most suitable at programming level.

Multithreading is preferred when you have to execute several independent tasks simultaneously. There is an inbuilt module named "threading" supported by Python, which is preferred for deploying the threads. There is a main thread available in all the Python programs. It is most suitable at programmatic level.

Program: Write a program to display the names of currently processing threads.

```
import threading
print("Current Executing Thread:",threading.current_thread().getName())

Current Executing Thread: MainThread
```

13.2 Ways of Creating a Thread

There are three ways of creating a thread:

13.2.1 Creation of a Thread without Using any Class

If the numbers of threads in the program are available, then we can't predict the order of execution and thus we can't predict the exact output for multithreaded programs. Hence, the exact output cannot be provided for the same program.

Example:

```python
from threading import *
def m1():
    for i in range(1,5):
        print("Child Thread Part")
t=Thread(target=m1)
t.start()
for i in range(1,5):
    print("Main Thread Part")
```

```
Child Thread PartMain Thread Part
Main Thread Part
Main Thread Part
Main Thread Part

Child Thread Part
Child Thread Part
Child Thread Part
```

13.2.2 Creation of a Thread by Extending the Thread Class

A child class needs to be created for the thread class. In the child class, the run() method needs to override the required task. On calling the start() method, the run() method automatically gets executed and will do the required task.

Example:

```python
from threading import *
class A(Thread):
    def run(self):
        for i in range(5):
            print("Child Thread Part")
a=A()
a.start()
for i in range(5):
    print("Main Thread Part")
```

```
Child Thread PartMain Thread Part
Main Thread Part
Main Thread Part
Main Thread Part
Main Thread Part

Child Thread Part
Child Thread Part
Child Thread Part
Child Thread Part
```

13.2.3 Creation of a Thread without Extending the Thread Class

Example:

```
from threading import *
class A:
    def m1(self):
        for i in range(5):
            print("Child Thread Part")
a=A()
t=Thread(target=a.m1)
t.start()
for i in range(5):
    print("Main Thread Part")
```

```
Child Thread PartMain Thread Part
Main Thread Part
Main Thread Part
Main Thread Part
Main Thread Part

Child Thread Part
Child Thread Part
Child Thread Part
Child Thread Part
```

Example (without multithreading):

```
from threading import *
import time
def m1(n):
    for i in n:
        time.sleep(1)
        print("Double:",2*i)
def m2(n):
    for i in n:
        time.sleep(1)
        print("Square:",i*i)
n=[2,4]
begintime=time.time()
m1(n)
m2(n)
print("Total time:",time.time()-begintime)
```

```
Double: 4
Double: 8
Square: 4
Square: 16
Total time: 4.0225300788879395
```

Example (with multithreading):

```
from threading import *
import time
def m1(n):
    for i in n:
        time.sleep(1)
        print("Double:",2*i)
def m2(n):
    for i in n:
        time.sleep(1)
        print("Square:",i*i)
n=[2,4]
begintime=time.time()
t1=Thread(target=m1,args=(n,))
t2=Thread(target=m2,args=(n,))
t1.start()
t2.start()
t1.join()
t2.join()
print("Total time:",time.time()-begintime)
```

```
Double:Square:  4
 4
Square:Double:   16
8
Total time: 2.0170021057128906
```

13.3 Setting and Getting Name of a Thread

In Python, all the threads have a name. The name can be a default name generated by PVM or the user-defined name assigned by the programmer. The following methods are used for getting and setting the name of the thread (Asad and Ali 2017):

- t.getName(): Name of the thread will be returned.
- t.setName(): Customized name will be set.

Example:

```
from threading import *
print(current_thread().getName())
current_thread().setName("Vijay")
print(current_thread().getName())
print(current_thread().name)
```

```
MainThread
Vijay
Vijay
```

13.4 Thread Identification Number

A uniquely identification number is provided by Python. This identification number can be accessed by the implicit variable "ident".

Example:

```
from threading import *
def A():
    print("Child Thread Part")
t=Thread(target=A)
t.start()
print("Main Thread Identification Number:",current_thread().ident)
print("Child Thread Identification Number:",t.ident)

Child Thread PartMain Thread Identification Number: 1940
Child Thread Identification Number: 4620
```

13.5 Active_count() Function

It returns the number of currently running active threads.

Example:

```
from threading import *
import time
def A():
    print(current_thread().getName(),"  start")
    time.sleep(3)
    print(current_thread().getName(),"  end")
    print("Active Threads:",active_count())
t1=Thread(target=A,name="ChildThread-1")
t2=Thread(target=A,name="ChildThread-2")
t1.start()
t2.start()
print("Active Threads:",active_count())
time.sleep(5)
print("Active Threads:",active_count())

ChildThread-1    start
ChildThread-2Active Threads: 7
    start
ChildThread-2ChildThread-1      end
Active Threads: 7
    end
Active Threads: 6
Active Threads: 5
```

13.6 Enumerate Function

It will produce a list of all the active threads that are presently running. It does not include the terminated threads or the threads that have not yet started.

Example:

```
from threading import *
import time
def m1():
    print(current_thread().getName(),"-->started")
    time.sleep(3)
    print(current_thread().getName(),"-->ended")
t1=Thread(target=m1,name="ChildThread-1")
t2=Thread(target=m1,name="ChildThread-2")
t1.start()
t2.start()
l=enumerate()
for t in l:
    print("Thread Name:",t.name)
time.sleep(5)
l=enumerate()
for t in l:
    print("Thread Name:",t.name)


ChildThread-1 -->started
ChildThread-2 -->started
Thread Name: MainThread
Thread Name: Thread-4
Thread Name: Thread-5
Thread Name: IPythonHistorySavingThread
Thread Name: Thread-3
Thread Name: ChildThread-1
Thread Name: ChildThread-2
ChildThread-1 -->ended
ChildThread-2 -->ended
Thread Name: MainThread
Thread Name: Thread-4
Thread Name: Thread-5
Thread Name: IPythonHistorySavingThread
Thread Name: Thread-3
```

13.7 IsAlive() Method

In Python, this is a built-in method of the thread class of the threading module. Using the thread object, it checks whether the specified thread is currently executing or not. If the thread is alive, it returns True else False.

Example:

```
from threading import *
import time
def m1():
    print(current_thread().getName(),"-->started")
    time.sleep(3)
    print(current_thread().getName(),"-->ended")
t1=Thread(target=m1,name="ChildThread-1")
t2=Thread(target=m1,name="ChildThread-2")
t1.start()
t2.start()
print(t1.name,"isAlive :",t1.isAlive())
print(t2.name,"isAlive :",t2.isAlive())
time.sleep(5)
print(t1.name,"isAlive :",t1.isAlive())
print(t2.name,"isAlive :",t2.isAlive())

ChildThread-1 -->started
ChildThread-2 -->started
ChildThread-1 isAlive : True
ChildThread-2 isAlive : True

ChildThread-1 -->ended
ChildThread-2 -->ended
ChildThread-1 isAlive : False
ChildThread-2 isAlive : False
```

13.8 Join() Method

If the thread wishes to wait until the completion of some other thread then the join() method is used.

By calling the join() method, the calling thread will be blocked until the main thread object gets terminated. The thread object gets terminated in one of the following cases:

- normal termination
- ill-handled exception
- until the optional timeout

Thus, the join() method shows the waiting state until the main thread object gets terminated. The time-out value can also be specified in association with the join method. It can be called any number of times.

Example: (Here, the main thread waits until the child thread completes its execution)

```
from threading import *
import time
def m1():
    for i in range(5):
        print("Vijay")
        time.sleep(2)
t=Thread(target=m1)
t.start()
t.join()
for i in range(3):
    print("Vimal")
```

```
Vijay
Vijay
Vijay
Vijay
Vijay
Vimal
Vimal
Vimal
```

Example: (Here, the main thread waits until the specified amount of time given in the join() method)

```
from threading import *
import time
def m1():
    for i in range(5):
        print("Vijay")
        time.sleep(2)
t=Thread(target=m1)
t.start()
t.join(5)
for i in range(3):
    print("Vimal")
```

```
Vijay
Vijay
Vijay
Vimal
Vimal
Vimal
Vijay
Vijay
```

13.9 Daemon Thread

Daemon threads are those that are executing at the background. They basically provide assistance for the non-daemon threads like garbage collector (GC). When the main thread faces low memory, then PVM will run the garbage collector instantly to remove the useless objects and to provide free memory, due to which the main thread continues the processing without facing memory issues.

Syntax (to check whether a thread is a daemon thread):   t.isDaemon()

Example:

```
from threading import *
print(current_thread().isDaemon())
print(current_thread().daemon)
```

```
False
False
```

The daemon nature can be modified by the setDaemon() method of the thread class.

Syntax: t.setDaemon(True)

Once the thread is started, we can't modify the daemon nature or an error will be raised

Example:

```
from threading import *
print(current_thread().isDaemon())
current_thread().setDaemon(True)
```

```
False
```

```
RuntimeError: cannot set daemon status of active thread
```

The daemon nature of the thread gets inherited from the parent to the child thread.

Default nature:
The default nature of the main thread is always non-daemon whereas for the rest of the threads the daemon nature gets inherited from parent to child, if the parent is a daemon thread then the child will also be a daemon thread and vice versa.

Example:

```
from threading import *
def m1():
    print("Swati")
t=Thread(target=m1)
print(t.isDaemon())
t.setDaemon(True)
print(t.isDaemon())
```

```
False
True
```

On the termination of the last non-daemon thread, all the daemon threads get terminated automatically. In the example given below, when the t.setDaemaon(True) is commented, then the main thread as well as the child thread are non-daemon in nature and thus both are executed until their completion.

Example:

```
from threading import *
import time
def m1():
    for i in range(5):
        print("Child Thread-1")
time.sleep(2)
t=Thread(target=m1)
#t.setDaemon(True)
t.start()
time.sleep(5)
print("End Of Main Thread")
```

```
Child Thread-1
Child Thread-1
Child Thread-1
Child Thread-1
Child Thread-1
End Of Main Thread
```

In the example given below, when the t.setDaemaon(True) is not commented, then the main thread acts as a non-daemon thread whereas the child thread are of daemon type and thus when the main thread terminates, the child thread automatically terminates.

Example:

```
from threading import *
import time
def m1():
    for i in range(5):
        print("Child Thread-1")
time.sleep(2)
t=Thread(target=m1)
t.setDaemon(True)
t.start()
time.sleep(5)
print("End Of Main Thread")
```

```
Child Thread-1
Child Thread-1
Child Thread-1
Child Thread-1
Child Thread-1
End Of Main Thread
```

13.10 Conclusion

After studying this chapter, the readers will be fluent with multithreading in Python. It can be summarized that multithreading can be used in cases where you would like to perform multiple input output bound tasks under the same application accessing shared resources.

Review Questions

1. When does the thread get terminated in the case of the join() method?
2. Explain the functions of a thread module.
3. Explain the thread objects and its methods.

Programming Assignments

PA 1. Write some Python code to illustrate the threading concept.

PA 2. Write a Python program to define a subclass using threading and instantiate the subclass and trigger the thread.

Reference

Asad, Ali, and Hamza Ali. 2017. *The C# programmer's study guide (MCSD): Exam.* Apress: 70–483.

14

Synchronization

14.1 Introduction

If a number of threads are executed at the same time then the data inconsistency problem may arise. In the below program, an irregular output is received as both the threads are executing together. In order to avoid this issue, synchronization should be used. Synchronization allows only one thread to execute at a time in order to overcome the data inconsistency problem. Synchronization is a mechanism that guarantees that if there is more than one thread that needs to be executed then a single thread at a time is permitted to access the critical section. The critical section is an important part of the program in which the shared resources are accessed. While ensuring synchronization primitives, two major issues such as the deadlock and race condition may arise.

Example:

```
from threading import *
import time
l=Lock()
def m1(n):
    l.acquire()
    for i in range(2):
        print("Hi:",end=' ')
        time.sleep(2)
        print(n)
    l.release()
t1=Thread(target=m1,args=("Parth",))
t2=Thread(target=m1,args=("Swati",))
t1.start()
t2.start()
```

```
Hi:Parth
Hi:Parth
Hi:Swati
Hi:Swati
```

The following techniques are used for implementing synchronization:

DOI: 10.1201/9781003185505-14

14.1.1 By Using the Lock Concept

The primitive lock is a synchronization primitive that is not governed by a specific thread once locked. This is the lowest level of synchronization primitive in Python, which is implemented by the thread extension module. There are two defined states of primitive lock:one is the locked state and other is the unlocked state. Its creation occurs in the unlocked state. Its two supported methods are acquire() and release(). When the state is unlocked, the acquire() method modifies the state as an unlocked state and immediately returns. In case of the locked state, the acquire() method blocks until the call to release() method in the other thread modifies it to unlocked state, then the acquire() method call resets it to locked state and returns (Sebesta 2012). The release() method is called only in the locked state; it modifies the state as unlocked state and immediately returns. If an attempt is made to release the unlocked lock, an error message will appear.

Example:

```
from threading import *
import time
def m1(n):
    for i in range(3):
        print("Hi:",end='')
        time.sleep(2)
        print(n)
t1=Thread(target=m1,args=("Vijay",))
t2=Thread(target=m1,args=("Swati",))
t1.start()
t2.start()
```

```
Hi:Hi:Swati
Hi:Vijay
Hi:Swati
Hi:Vijay
Hi:Swati
Vijay
```

Example:

```
from threading import *
l = Lock()
#L.acquire()
l.release()
```

```
RuntimeError: release unlocked lock
```

Simple lock:

The standard lock object doesn't consider which thread is presently holding the lock. The thread that is attempting to acquire the lock will be blocked if the lock is being held by another thread, even if the same thread is holding that lock in advance.

Example:

```
from threading import *
l=Lock()
print("Acquire Lock")
l.acquire()
print("Acquire Lock Again")
l.acquire()
```

```
Acquire Lock
Acquire Lock Again
```

If the thread recursively accesses the resources, then the thread acquires the same lock again, which might lead to blocking of the thread. Thus, this old method of locking is not suitable for the execution of recursive functions (Satyanarayana 2020). Reentrant locks are preferred for dealing with such problems.

Example:

```
from threading import *
l=RLock()
print("Acquire Lock")
l.acquire()
print("Acquire Lock Again")
l.acquire()
```

14.1.2 By Using RLock

RLock is the reentrant lock. It is the synchronization primitive that can be acquired a number of times by the same thread. It follows the concepts of owning the thread and recursion level in addition to the locked or unlocked state used by primitive locks. In the case of a locked state, some of the threads own the lock whereas in the case of an unlocked state, no thread owns the lock. The thread calls the acquire() method, for locking the lock. This returns once the thread owns the lock. The thread calls the release() method to unlock the lock.

Example:

```
from threading import *
import time
l=RLock()
def fact(n):
    l.acquire()
    if n==0:
        opt=1
    else:
        opt=n*fact(n-1)
    l.release()
    return opt
def opt(n):
    print("Factorial",n,":",fact(n))
t1=Thread(target=opt,args=(3,))
t2=Thread(target=opt,args=(5,))
t1.start()
t2.start()
```

```
Factorial 3 : 6
Factorial 5 : 120
```

Table 14.1 shows the difference between RLock and Lock.

TABLE 14.1
Lock vs RLock

Lock	Rlock
Only one thread can acquire it at a time including the main thread.	Only one thread can acquire it at a time, but the main thread can acquire it any number of times.
It is not preferred in recursive and nested functions.	It is most suitable in recursive and nested functions.
It is responsible only for locking and unlocking of objects; it doesn't take care of the main thread and recursion level.	It is responsible for lock, unlock, main thread and recursion level as well.

14.1.3 By Using Semaphore

Semaphores are preferred to constrain the access of the shared resources. The semaphore takes care of the internal counter which is decremented by all the acquire() calls and incremented by all the release() call. The counter cannot be less than zero. The counter denotes the maximum quantity of threads that are accessed at the same time. Its default value is 1.

Syntax (creation of semaphore object):s= Semaphore (counter)

Case 1:   s = Semaphore()

Here, the counter value is 1 and access is permitted to only one thread at a time.

Case2:   s=Semaphore(3)

Here, a semaphore object can be accessed by three threads simultaneously. The remaining threads must wait for the semaphore to be released.

Example:

```
from threading import *
import time
s=Semaphore(2)
def m1(n):
    s.acquire()
    for i in range(4):
        print("Hi:",end='')
        time.sleep(2)
        print(n)
    s.release()
t1=Thread(target=m1,args=("Vimal",))
t2=Thread(target=m1,args=("Deepak",))
t1.start()
t2.start()
```

```
Hi:Hi:DeepakVimal
Hi:
Hi:DeepakVimal

Hi:Hi:DeepakVimal
Hi:
Hi:VimalDeepak
```

Bounded semaphore:
There is no limit on general semaphores. The release() method can be called 'n' number of times for incrementing the value of counter. It can even cross the number of acquire() calls.

Example:

```
from threading import *
s=Semaphore(2)
s.acquire()
s.acquire()
s.release()
s.release()
s.release()
s.release()
print("End")
```

```
End
```

A bounded semaphore is similar to a semaphore with the only difference that the number of release() calls has to be less than the number of acquire() calls or an error message will be raised.

Example:

```
from threading import *
s=BoundedSemaphore(2)
s.acquire()
s.acquire()
s.release()
s.release()
s.release()
s.release()
print("End")
```

```
ValueError: Semaphore released too many times
```

It is not valid because the quantity of release() calls has exceeded the quantity of acquire() calls in the bounded semaphore. The basic difference between a lock and a semaphore object is that the lock object can be acquired by a single thread only whereas the semaphore object can be acquired by the number of threads mentioned in the counter value.

14.2 Inter-Thread Communication

During the execution of the program, threads are required to interact with each other, which is known as inter-thread communication. For example, in a producer-consumer problem, after producing an item, the producer has to signal to the consumer that the item is produced so that the consumer can consume. In Python, inter-thread communication can be implemented as follows:

14.2.1 By Using Event Objects

The simplest mechanism for inter-thread communication is the event object. One of the threads will signal and the other thread will wait for it to receive.

Syntax of creation: event = threading.Event()

The methods of the event_class are discussed below:

1. set(): GREEN signal color will be used for all the waiting threads whereas the Boolean value of an internal flag will be True.
2. clear(): RED signal color will be used for all the waiting threads whereas the Boolean value of an internal flag will be False.
3. isSet(): It is used to check whether the specified event is set or not.
4. wait() or wait(seconds): Threads will wait until the events are set.

Example:

```
from threading import *
import time
def producer():
    time.sleep(3)
    print("Producer thread producing items:")
    print("Producer thread notification")
    event.set()
def consumer():
    print("Consumer thread waiting")
    event.wait()
    print("Consumer thread consuming items")
event=Event()
t1=Thread(target=producer)
t2=Thread(target=consumer)
t1.start()
t2.start()
```

```
Consumer thread waiting
Producer thread producing items:
Producer thread notification
Consumer thread consuming items
```

14.2.2 By Using a Condition Object

Another method for inter-thread communication is the condition object. The condition indicates that there is a change in the state such as the producer has produced an item or the consumer has consumed an item. The threads will wait for that condition to occur and they will be informed after the occurrence. The condition is defined with respect to the lock. There are two methods that the condition comprises, i.e. the acquire() method and release() method, which will
call the method of the respective lock.

Syntax (creation of condition object):condition = threading.Condition()

The different methods of condition object are discussed below:

1) acquire(): It is used to obtain the condition object prior to producing an item or consuming an item, i.e. the threads will acquire the internal locks.
2) release(): It is used to give away the condition object after producing an item or consuming an item; the threads will give away the internal lock.
3) wait() or wait(time): The threads will wait until they get any notification or time expiry.
4) notify(): The notification will be sent to one of the waiting threads.
5) notifyAll(): The notification will be sent to all the waiting threads.

Case study of producer-consumer problem:
The producing thread of the producer needs to acquire the condition before producing an item to the resource and signaling consumers whereas the consumer must acquire the condition and then it consumes items from the resource.

Example:

```
from threading import *
def consume(c):
    c.acquire()
    print("Consumer waiting")
    c.wait()
    print("Consumer notification")
    c.release()
def produce(c):
    c.acquire()
    print("Producer Producing Items")
    print("Producer Notification")
    c.notify()
    c.release()
c=Condition()
t1=Thread(target=consume,args=(c,))
t2=Thread(target=produce,args=(c,))
t1.start()
t2.start()
```

```
Consumer waiting
Producer Producing Items
Producer Notification
Consumer notification
```

Example:

```
from threading import *
import time
import random
items=[]
def produce(c):
    while True:
        c.acquire()
        item=random.randint(1,25)
        print("Producer Producing Item:",item)
        items.append(item)
        print("Producer Notification")
        c.notify()
        c.release()
        time.sleep(3)
def consume(c):
    while True:
        c.acquire()
        print("Consumer waiting")
        c.wait()
        print("Consumer consumed the item",items.pop())
        c.release()
        time.sleep(3)
c=Condition()
t1=Thread(target=consume,args=(c,))
t2=Thread(target=produce,args=(c,))
t1.start()
t2.start()
```

```
Consumer waiting
Producer Producing Item: 21
Producer Notification
Consumer consumed the item 21
```

and so on...

14.2.3 By Using Queue

The final method for inter-thread communication is by queues. It is the most preferred method for inter-thread communication. Internally, the queue has its condition with the associated lock. Thus, while working with queues, one does not have to think about synchronization.

Syntax (to import queue module): import queue
Syntax (creation of queue object): q = queue.Queue()

The queue methods are defined below:

- put(): To put an item in the queue.
- get(): To remove an item and return an item from the queue.

The producer thread uses the put() method to add a data item into the queue. After adding a data item, the lock will automatically be released. The put() method also verifies whether the queue is full or not and if it finds the queue is full then the producer thread will enter the waiting state by calling for the wait() method. The consumer thread uses the get() method to remove and get data from the queue. Once removal is completed then the lock automatically gets released. If the queue is empty then the consumer thread will enter the waiting state by calling the wait() method. Once the queue is updated with data then the thread automatically gets notified.

Example:

```
from threading import *
import time
import random
import queue
def produce(q):
    while True:
        item=random.randint(1,25)
        print("Producer Producing Item:",item)
        q.put(item)
        print("Producer Notification")
        time.sleep(3)
def consume(q):
    while True:
        print("Consumer waiting ")
        print("Consumer consumed the item:",q.get())
        time.sleep(3)
q=queue.Queue()
t1=Thread(target=consume,args=(q,))
t2=Thread(target=produce,args=(q,))
t1.start()
t2.start()

Consumer waiting
Producer Producing Item: 6
Producer Notification
Consumer consumed the item: 6
```
and so on...

14.3 Variants of Queues

Python provides three variants of queues discussed below:

14.3.1 First-In-First-Out Queue

As the name suggests, the item first inserted in the queue will be the first to be removed from the queue.

> q = queue.Queue()

Example:

```
import queue
q=queue.Queue()
q.put(12)
q.put(15)
q.put(25)
q.put(45)
while not q.empty():
    print(q.get(),end=' ')
```

```
12 15 25 45
```

14.3.2 Last-In-First-Out Queue

As the name suggests, the item first inserted in the queue will be the last to be removed from the queue.

Example:

```
import queue
q=queue.LifoQueue()
q.put(12)
q.put(15)
q.put(25)
q.put(45)
while not q.empty():
    print(q.get(),end=' ')
```

```
45 25 15 12
```

14.3.3 Priority Queue

As the name suggests, the item will be inserted on the basis of some priority assigned to it.

Example:

```
import queue
q=queue.PriorityQueue()
q.put(25)
q.put(12)
q.put(9)
q.put(18)
while not q.empty():
    print(q.get(),end=' ')
```

```
9 12 18 25
```

In the case of non-numeric data, tuples are used (p,q) where p is the priority and q is the element.

Example:

```
import queue
q=queue.PriorityQueue()
q.put((4,"vijay"))
q.put((1,"swati"))
q.put((2,"shiv"))
q.put((3,"deepak"))
while not q.empty():
    print(q.get()[1],end=' ')
```

```
swati shiv deepak vijay
```

14.4 Usage of Locks

The important cases are:

Case 1:
It is strongly suggested to use the finally block to write the lock releasing code. The key point here is that whether the exception occurs or not, the lock will always be released.

l = threading.Lock()
l.acquire()
try:
    required to perform safe operation
finally:
    l.release()

Example:

```
from threading import *
import time
l=Lock()
def m1(n):
    l.acquire()
    try:
        for i in range(3):
            print("Hi:",end='')
            time.sleep(2)
            print(n)
    finally:
        l.release()
t1=Thread(target=m1,args=("Vijay",))
t2=Thread(target=m1,args=("Deepak",))
t1.start()
t2.start()
```

```
Hi:Vijay
Hi:Vijay
Hi:Vijay
Hi:Deepak
Hi:Deepak
Hi:Deepak
```

Case 2:
It is strongly suggested to use the with statement for acquiring the lock. The key point here is that upon reaching the end of the with block, the lock will automatically be released.

For file:
    with open('abc.txt','w') as f:
            f.write("Python")

For Lock:
            lock = threading.Lock()
            with lock:
            required to perform safe operation
            automatically the lock gets release

Example:

```
from threading import *
import time
lock=Lock()
def m1(n):
    with lock:
        for i in range(3):
            print("Hi:",end='')
            time.sleep(2)
            print(n)
t1=Thread(target=m1,args=("Deepak",))
t2=Thread(target=m1,args=("Shiv",))
t1.start()
t2.start()
```

```
Hi:Deepak
Hi:Deepak
Hi:Deepak
Hi:Shiv
Hi:Shiv
Hi:Shiv
```

14.5 Conclusion

In this chapter, you have seen different aspects of Python threading along with many examples to build the threaded programs and the issues they resolve. You have also seen the instances of the problems that arise while writing and debugging threaded programs.

Review Questions

1. Mention the benefit of with statements to acquire a lock in threading.
2. Explain timer objects.
3. Explain condition objects.
4. What are RLock objects?

Programming Assignments

PA 1. Write some Python code showing different cases of lock.
PA 2. Write some Python code showing different cases of RLock.

References

Satyanarayana, M. S. 2020. Language fundamentals on Python. A promotional document by Durga
        Software Solutions uploaded as open source document. https://doku.pub/documents/
        python-durgapdf-8lyrmkdrne0d (accessed January 09, 2021)
Sebesta, Robert W. 2012. *Concepts of programming languages*. Pearson.

15

Regular Expressions and Web Scraping

15.1 Introduction

Regular expressions are basically used if you want to generate a particular pattern or format from a set of strings such as representation of all the mobile numbers or the e-mail IDs using regular expression. The general areas of using regular expression are:

- to deploy the validation frameworks
- to deploy pattern matching applications
- to develop translators like compiler, interpreter, etc.
- to develop various communication protocols
- to develop digital circuits.

Module to use regular-expressions: re

The re module comprises various inbuilt functions for using regular expressions in applications.

(i) compile(): It is used for compiling the pattern into RegexObject.

Syntax: pattern = re.compile("pqr")

(ii) finditer(): It produces the iterator object which returns the match object for all the matches.

Syntax: match = pattern.finditer("pqpqpqqppp")

There are a number of methods that can be called by the match object, which are defined in Table 15.1.

TABLE 15.1
Methods of match object

Method	Function
start()	Start_index of match will be returned
end()	(end_index+1) of match will be returned
group()	matched string will be returned

DOI: 10.1201/9781003185505-15

Example:

```
import re
count=0
p=re.compile("pq")
m=p.finditer("pqppqpq")
for m1 in m:
    count+=1
    print(m1.start(),"-->",m1.end(),"-->",m1.group())
print("Number of occurrences: ",count)
```

```
0 --> 2 --> pq
3 --> 5 --> pq
5 --> 7 --> pq
Number of occurrences:  3
```

Example: (patterns passed as argument)

```
import re
count=0
m=re.finditer("pq","pqppqpq")
for m1 in m:
    count+=1
    print(m1.start(),"-->",m1.end(),"-->",m1.group())
print("Number of occurrences: ",count)
```

```
0 --> 2 --> pq
3 --> 5 --> pq
5 --> 7 --> pq
Number of occurrences:  3
```

15.1.1 Character Classes

Different character classes are used for searching a group of characters. They are defined in Table 15.2.

TABLE 15.2

Character classes

Class	Function
[pqr]	Either 'p' or ' q' or 'r'
[^pqr]	Except p and q and r
[a-z]	Any alphabet in lower case
[A-Z]	Any alphabet in upper case
[a-zA-Z]	Any lower or upper case alphabet
[0-9]	Any digit from 0, 1, 2, 3, 4, 5, 6, 7, 8, 9
[a-zA-Z0-9]	Any alpha-numeric character
[a-zA-Z0-9]	Any alpha-numeric character
[^a-zA-Z0-9]	Except alpha-numeric characters

Example:

```
import re
m=re.finditer("[adk]","a5d@k8z")
for m1 in m:
    print(m1.start(),"-->",m1.group())

0 --> a
2 --> d
4 --> k
```

Example:

```
import re
m=re.finditer("[^adk]","a5d@k8z")
for m1 in m:
    print(m1.start(),"-->",m1.group())

1 --> 5
3 --> @
5 --> 8
6 --> z
```

Example:

```
import re
m=re.finditer("[d-p]","a5d@k8mz")
for m1 in m:
    print(m1.start(),"-->",m1.group())

2 --> d
4 --> k
6 --> m
```

Example:

```
import re
m=re.finditer("[1-8]","a5d@6k8mz")
for m1 in m:
    print(m1.start(),"-->",m1.group())

1 --> 5
4 --> 6
6 --> 8
```

Example:

```
import re
m=re.finditer("[a-zA-Z0-9]","a5d@6k8mz")
for m1 in m:
    print(m1.start(),"-->",m1.group())

0 --> a
1 --> 5
2 --> d
4 --> 6
5 --> k
6 --> 8
7 --> m
8 --> z
```

Example:

```
import re
m=re.finditer("[^a-zA-Z0-9]","a5d@6k#8mz")
for m1 in m:
    print(m1.start(),"-->",m1.group())
```

```
3 --> @
6 --> #
```

15.1.2 Pre-defined Character Classes

The types of pre-defined character classes are defined in Table 15.3.

TABLE 15.3
Predefined character class

Class	Function
\d	Any digit from 0, 1, 2, 3, 4, 5, 6, 7, 8, 9
\D	Any character except the digits
\s	Space Character
\S	Any character except the space character
.	Any character including the special character
\w	Any word character [a-zA-Z0-9]
\W	Any character except the word character

Example:

```
import re
m=re.finditer("\S","a5d@6k#8mz")
for m1 in m:
    print(m1.start(),"-->",m1.group())
```

```
0 --> a
1 --> 5
2 --> d
3 --> @
4 --> 6
5 --> k
6 --> #
7 --> 8
8 --> m
9 --> z
```

Example:

```
import re
m=re.finditer("\d","a5d@6k#8mz")
for m1 in m:
    print(m1.start(),"-->",m1.group())
```

```
1 --> 5
4 --> 6
7 --> 8
```

Example:

```
import re
m=re.finditer("\D","a5d@6k#8mz")
for m1 in m:
    print(m1.start(),"-->",m1.group())
```

```
0 --> a
2 --> d
3 --> @
5 --> k
6 --> #
8 --> m
9 --> z
```

Example:

```
import re
m=re.finditer("\w","a5d@6k#8mz")
for m1 in m:
    print(m1.start(),"-->",m1.group())
```

```
0 --> a
1 --> 5
2 --> d
4 --> 6
5 --> k
7 --> 8
8 --> m
9 --> z
```

Example:

```
import re
m=re.finditer("\W","a5d@6k#8mz")
for m1 in m:
    print(m1.start(),"-->",m1.group())
```

```
3 --> @
6 --> #
```

Example:

```
import re
m=re.finditer(".","a5d@6k#8mz")
for m1 in m:
    print(m1.start(),"-->",m1.group())
```

```
0 --> a
1 --> 5
2 --> d
3 --> @
4 --> 6
5 --> k
6 --> #
7 --> 8
8 --> m
9 --> z
```

Example:

```
import re
m=re.finditer("\s","a5d@ 6k#8 mz")
for m1 in m:
    print(m1.start(),"-->",m1.group())

4 -->
9 -->
```

15.1.3 Quantifiers

The quantifiers are used for specifying the number of occurrences to match. Table 15.4 shows the types of quantifiers and their associated functions (Gajda 2015).

TABLE 15.4
Types of quantifiers

Quantifier	Function
p	Exactly one 'p'
p+	At least one 'p'
p*	Any number of 'p's including the zero number
p{k}	Exactly k number of 'p's
p?	At most one 'p' either zero or one
p{k,r}	Minimum k number of 'p's and Maximum r number of 'k's

Example:

```
import re
m=re.finditer("p","pqppqqppqq")
for m1 in m:
    print(m1.start(),"-->",m1.group())

0 --> p
2 --> p
3 --> p
6 --> p
7 --> p
```

Example:

```
import re
m=re.finditer("p+","pqppqqppqq")
for m1 in m:
    print(m1.start(),"-->",m1.group())

0 --> p
2 --> pp
6 --> pp
```

Example:

```
import re
m=re.finditer("p*","pqppqqppqq")
for m1 in m:
    print(m1.start(),"-->",m1.group())
```

```
0 --> p
1 -->
2 --> pp
4 -->
5 -->
6 --> pp
8 -->
9 -->
10 -->
```

Example:

```
import re
m=re.finditer("p?","pqppqqppqq")
for m1 in m:
    print(m1.start(),"-->",m1.group())
```

```
0 --> p
1 -->
2 --> p
3 --> p
4 -->
5 -->
6 --> p
7 --> p
8 -->
9 -->
10 -->
```

Example:

```
import re
m=re.finditer("q{4}","pqppqqppqqqqqq")
for m1 in m:
    print(m1.start(),"-->",m1.group())
```

```
8 --> qqqq
```

Example:

```
import re
m=re.finditer("q{2,3}","pqppqqppqqqqqq")
for m1 in m:
    print(m1.start(),"-->",m1.group())
```

```
4 --> qq
8 --> qqq
11 --> qqq
```

15.2 Functions of Re Module

15.2.1 Match()

The purpose of the match function is to match a regular expression pattern to a string with optional flags. To use this function, the 're' module needs to be imported. It will return a match object if the matching was a success and returns None if the matching was a failure.

Syntax: re.match (pattrn, str, flags = 0)

where, pattrn is the regular expression that is to be matched
              str is searched to match the pattern at the beginning of string
              flags are modifiers specified using bitwise OR

Example:

```
import re
s=input("Enter pattern: ")
m=re.match(s,"pqrstuvw")
if m!= None:
    print("Match Beginning")
    print("Start Index:",m.start(), " End Index:",m.end())
else:
    print("Not Match")
```
```
Enter pattern: pqrstu
Match Beginning
Start Index: 0   End Index: 6

Enter pattern: abc
Not Match
```

15.2.2 Fullmatch()

The fullmatch() function is used to match a pattern to the entire target string; the full string has to be matched in accordance with the given pattern. If the full string matches then it will return the match object or None will be returned.

Example:

```
import re
s=input("Enter pattern: ")
m=re.fullmatch(s,"pqpqp")
if m!= None:
    print("Full Matched")
else:
    print("Not full Matched")
```
```
Enter pattern: pqpqp
Full Matched

Enter pattern: pq
Not full Matched
```

15.2.3 Search()

This function searches for the first occurrence of the regular expression pattern within a string with the optional flags. The match object will be returned if the matching occurs else None will be returned.

Syntax: re.search(pattrn, str, flags =0)

where, pattrn, str and flags have the same meaning as that of match() function.

Example:

```
import re
s=input("Enter pattern :  ")
m=re.search(s,"pqppppqqp")
if m!= None:
    print("Match Pattern")
    print("Start index:",m.start(),"End index:",m.end())
else:
    print("Not Match")
```

```
Enter pattern : pppp
Match Pattern
Start index:  2 End index:  6

Enter pattern : stp
Not Match
```

15.2.4 Findall()

If you wants to find all the occurrences of the match, then the findall() function is used. It produces a list object containing all the occurrences.

Example:

```
import re
l=re.findall("[0-9]","b5c9d4pz1")
print(l)
```

```
['5', '9', '4', '1']
```

15.2.5 Finditer()

This produces an iterator that gives a match object for the entire match. For all match objects start(), end() and group() functions can be called.

Example:

```
import re
i=re.finditer("[a-z]","b5c9d4pz1")
for m in i:
    print(m.start(),"-->",m.end(),"-->",m.group())
```

```
0 --> 1 --> b
2 --> 3 --> c
4 --> 5 --> d
6 --> 7 --> p
7 --> 8 --> z
```

15.2.6 Sub()

Here, sub denotes the substitution or the replacement. The matched patterns will be replaced by the provided replacement, in the target_string.

Syntax: re.sub(regex,replacement,target_string)

Example:

```
import re
s=re.sub("[a-z]","$","b5c9d4pz1")
print(s)
```

```
$5$9$4$$1
```

15.2.7 Subn()

The subn() function is quite similar to the sub() function, except the point that it returns the number of replacement. It gives a tuple in which the primary element is the resultant string and the secondary element is the number of replacement.

Syntax: (result_string, number-of-replacements)

Example:

```
import re
s=re.subn("[a-z]","$","b5c9d4pz1")
print(s)
print("String:",s[0])
print("Replacements:",s[1])
```

```
('$5$9$4$$1', 5)
String: $5$9$4$$1
Replacements: 5
```

15.2.8 Split()

The split() function is useful when you want to divide the specified string in accordance with the particular pattern. It gives the list of all tokens.

Example:

```
import re
s=re.split(",","vijay,swati,vimal,shashwat")
print(s)
for i in s:
    print(i)
```

```
['vijay', 'swati', 'vimal', 'shashwat']
vijay
swati
vimal
shashwat
```

Example:

```
import re
l=re.split("\.","www.google.com")
for t in l:
    print(t)
```

```
www
google
com
```

15.2.9 ^ Symbol

The symbol ^ is used for checking whether the specified string starts with the mentioned pattern or didn't match. If the target string matches with the mentioned pattern, the match object will be returned or None will be returned.

Example:

```
import re
s="Dr Kalam is President of India"
r=re.search("^Dr",s)
if r != None:
    print("Start by Dr")
else:
    print("Not start by Dr")
```

```
Start by Dr
```

15.2.10 $ Symbol

The $ symbol is used for checking whether the specified string ends with the mentioned pattern or didn't match. If the specified string ends with the mentioned pattern, it will give the match object or None will be returned.

Example:

```
import re
s="Dr Kalam is President of India"
r=re.search("India$",s)
if r != None:
    print("Ended by India")
else:
    print("Not ended by India")
```

```
Ended by India
```

If you wants to ignore the case, then you have to pass the third argument in the search() function as follows:

re.IGNORECASE

Example:

```
import re
s="Dr Kalam is President of India"
r=re.search("india$",s,re.IGNORECASE)
if r != None:
    print("Ended by India")
else:
    print("Not ended by India")
```

```
Ended by India
```

Program: Write some code in Python to identify whether the entered mobile number is valid or not.

```
import re
n=input("Enter number:")
m=re.fullmatch("[7-9]\d{9}",n)
if m!= None:
    print("Valid Cell Number")
else:
    print("Invalid Cell Number")
```

```
Enter number:7896785644
Valid Cell Number
```

```
Enter number:9835
Invalid Cell Number
```

Program: Write some code in Python to extract all the mobile numbers from a text file into another text file.

```
import re
f1=open("abc.txt","r")
f2=open("xyz.txt","w")
for l in f1:
    l1=re.findall("[7-9]\d{9}",l)
    for n in l1:
        f2.write(n+"\n")
        print("Extracted all Contact Numbers into xyz.txt")
f1.close()
f2.close()
```

```
Extracted all Contact Numbers into xyz.txt
Extracted all Contact Numbers into xyz.txt
Extracted all Contact Numbers into xyz.txt
```

15.3 Web Scraping

Web scraping is the procedure of collecting the information from different web pages. It is an automated procedure that is used to extricate the massive amount of data from various websites. Generally, the data available on websites is in unstructured form. Web scraping is used to gather the available unstructured form of data and convert it in the structured form (Naidu 2020). There are numerous applications of web scraping like collection of email addresses, social media scraping, research and development, and comparison of prices of any article, etc.

Example:

```python
import re,urllib
import urllib.request
s="google scholar".split()
print(s)
for s1 in s:
    print("Search",s1)
    u=urllib.request.urlopen("http://"+s1+".com")
    t=u.read()
    title=re.findall("<title>.*</title>",str(t),re.I)
    print(title[0])
```

```
['google', 'scholar']
Search google
<title>Google</title>
Search scholar
```

Program: Write some Python code to extract all mnnit.ac.in phone numbers by using regular expressions and web scraping.

```python
import re,urllib
import urllib.request
u=urllib.request.urlopen("http://www.mnnit.ac.in/index.php/contact-us")
t=u.read()
num=re.findall("[0-9-]{7}[0-9-]+",str(t),re.I)
for n in num:
    print(n)
```

```
2016-10-27-05-34-33
11163860651212808366 9
6163079620322188018
11163860651212808366 9
6077814256575229809
2016-10-27-05-34-33
11163860651212808366 9
6163079620322188018
11163860651212808366 9
6077814256575229809
2018-05-07-10-53-20
-2018-19
-2016-17
2018-11-05
91-0532-2545404
91-0532-2545341
```

Program: Write some code in Python to identify whether the entered e-mail ID is valid/invalid.

```python
import re
s=input("Enter E-Mails:")
m=re.fullmatch("\w[a-zA-Z0-9_.]*@gmail[.]com",s)
if m!=None:
    print("Valid");
else:
    print("Invalid")
```

```
Enter E-Mails:vijay@gmail.com
Valid

Enter E-Mails:vijaygmail.com
Invalid
```

Program: Write some code in Python to identify whether the entered vehicle number is valid or invalid.

```python
import re
s=input("Enter Vehicle Number:")
m=re.fullmatch("UP[019][0-9][A-Z]{2}\d{4}",s)
if m!=None:
    print("Valid Vehicle Number");
else:
    print("Invalid Vehicle Number")
```

```
Enter Vehicle Number:UP15AX6460
Valid Vehicle Number

Enter Vehicle Number:UP156460
Invalid Vehicle Number
```

15.4 Conclusion

Regular expressions are crucial to all software developers and after studying this chapter, you can see how easily you can use regular expressions in Python. It has been seen that with RegEx, pattern matching can be used for searching particular strings of characters rather than constructing multiple, literal search queries. Different operations in Python's re module are included and the method to use it in your personal Python application is explained. Finally, an important application of regular expression known as web scraping has been explained with a sufficient number of examples.

Review Questions

1. Mention the purpose of the match() function in regular expressions.
2. Explain the findall() function.
3. Mention the purpose of the search() function in regular expressions.
4. Differentiate between replace() and search() in regular expressions.
5. Explain the significance of the re module in regular expressions.

Programming Assignment

PA 1: Write a regular expression to denote all the ten digit mobile numbers.

PA 2: Create a file named abc.txt. and retrieve all the lines that contain "A".

PA 3: Create a file named abc.txt. and retrieve an odd digit followed by the even digit.

PA 4: Create a file named abc.txt. and retrieve the lines with the word "the" one or more times.

References

Gajda, Włodzimierz. 2015. Default configuration and security settings of the guest VM. In *Pro Vagrant*. Apress: 65–83.

Naidu, C. 2020. Language fundamentals on Python. Notes of Python programming language document by Durga Software Solutions uploaded as open source document. www.scribd.com/document/427237165/Python-notes (accessed February 03, 2021)

Satyanarayana, M. S. 2020. Language fundamentals on Python. A promotional document by Durga Software Solutions uploaded as an open source document. https://doku.pub/documents/python-durgapdf-8lyrmkdrne0d (accessed February 03, 2021)

16

Database Programming

16.1 Introduction

On the basis of our requirements, we need to save our data such as office data, accounts data, student data, and faculty data and so on. In order to store such data, storage areas are required. Basically, there are two types of storage areas:

- Permanent storage area: They are used to store the data permanently like the file system, data warehouses, data clouds, etc.
- Temporary storage area: They are used to store the data temporarily like lists or a dictionary of Python objects.

16.2 File System

File systems are provided by the local operating system. They are preferred for storing smaller amounts of data. The constraints of a file system are:

- Huge amount of data cannot be stored
- Query language is not supported
- No security to data
- Inconsistent data.

To overcome these file system problems, databases are preferred.

16.3 Database

With the use of databases, huge amounts of information can be stored. Query language support is available for all the databases. In order to make the data secure, usernames and passwords are required for accessing the database. The data is stored in the form of tables. While designing the schema of tables, the database administrator follows different

DOI: 10.1201/9781003185505-16

normalizations and can inherit different constraints such as null key, primary key, and so on, which prevents the duplication of data (Stones and Matthew2006). Thus, data inconsistency will not occur.

The limitations of databases are:

- Can't hold big data like terabytes or gigabytes of data
- No support for unstructured (like audio, video, etc.) and semi-structured data (like xml files).

To overcome these database problems, advanced storage areas such as data warehouses and big data technologies such as Hadoop are used.

16.3.1 Python Database Programming

According to the programming need, sometimes the programmer needs to connect with the database and some operations needs to be performed such as the creation of a table, insertion of data, updating the data, deletion of data, selecting the data, and so on. Python can be used to send SQL commands to the database. Python assists support to many databases like MySQL, Oracle, GadFly, SQLserver etc. It comprises various modules for different databases.

The steps to be followed in Python database programming are as follows:

Step 1:
Import the database specific module like cx_Oracle for Oracle database.

Syntax: import cx_Oracle

Step 2:
Establish the connection between the database and the Python programs. For this, the connect() function is used in Oracle.

Syntax: conn = cx_Oracle.connect(database information)

Step 3:
Execution of SQL queries and cursor objects is required to hold their results. For this, cursor() object is used in Oracle.

Syntax: csr = conn.cursor()

Step 4:
Execution of SQL queries using the cursor object. Table 16.1 shows the methods used.

TABLE 16.1
Methods to execute SQL query using cursor object

S.NO.	Methods	Description
1	execute(sqlquery)	Execution of a single SQL query
2	executemany()	Execution of parameterized query
3	executescript(sqlqueries)	Execution of a string of SQL queries separated by semicolon.

Step 5:
In case of Data Manipulation Language (DML) queries, use the commit and rollback statement. The commit operation is used to save the changes in the database whereas the rollback operation is used to roll the temporary changes back to the prior state.

Step 6:
Use of cursor object for fetching the output of select query. Table 16.2 shows different queries and their associated descriptions.

TABLE 16.2
Fetch operations

S. No.	Queries	Description
1	fetchall()	Fetching all the rows and returning the list of rows
2	fetchone()	Fetching only a single row
3	fetchmany(n)	Fetching the first n rows

Step 7:
Close the resources after completion of operations. They are closed in the reverse order of opening of file. For this, the close() method is used.

Syntax: csr.close()   OR   conn.close()

The following are a few important methods used in Python database programming:

- connect()
- execute()
- cursor()
- executemany()
- executescript()
- rollback()
- commit()
- fetch
- fetchall()
- fetchone()
- fetchmany(n)
- close()

16.3.2 Working with the Oracle Database

If the programmer needs to communicate with the database using a Python program then some translators are required for translating the Python calls into the database specific calls and vice versa. This translator is known as a driver or connector.

Using help module:

```
help ('modules')
Crypto              brain_ssl           math                smtpd
CustomLogger        brain_subprocess    matplotlib          smtplib
Cython              brain_threading     mccabe              sndhdr
IPython             brain_typing        menuinst            snowballstemmer
OpenSSL             brain_uuid          mimetypes           socket
PIL                 bs4                 mistune             socketserver
PyQt5               builtins            mkl                 socks
__future__          bz2                 mkl_fft             sockshandler

-----------

brain_pytest        macpath             sipdistutils        zipfile
brain_qt            mailbox             site                zipimport
brain_random        mailcap             six                 zipp
brain_re            markupsafe          skimage             zlib
brain_six           marshal             sklearn             zmq

Enter any module name to get more help.  Or, type "modules spam" to search
for modules whose name or summary contain the string "spam".
```

Program: Write a code for connecting with the Oracle database.

```python
import cx_Oracle
conn=cx_Oracle.connect('scott/tiger@localhost')
print(conn.version)
conn.close()
```

Program: Write a program for creating a faculty table in the Oracle database.

Faculty(Fac_id, Fac_name, Fac_sal)

```python
import cx_Oracle
try:
    conn=cx_Oracle.connect('scott/tiger@localhost')
    csr=conn.cursor()
    csr.execute("create table Faculty(Fac_id number,
                Fac_name varchar2(8),Fac_sal number(8,2))")
    print("Create Table")
except cx_Oracle.DatabaseError as e:
    if conn:
        conn.rollback()
        print("Connectivity problem",e)
finally:
    if csr:
        csr.close()
    if conn:
        conn.close()
```

Program: Write some code to drop the faculty table.

```
import cx_Oracle
try:
    conn=cx_Oracle.connect('scott/tiger@localhost')
    csr=conn.cursor()
    csr.execute("drop table Faculty")
    print("Table dropped")
except cx_Oracle.DatabaseError as e:
    if conn:
        conn.rollback()
        print("Connectivity problem",e)
finally:
    if csr:
        csr.close()
    if conn:
        conn.close()
```

Program: Write some code for inserting a row in the faculty table.

```
import cx_Oracle
try:
    conn=cx_Oracle.connect('scott/tiger@localhost')
    csr=conn.cursor()
    csr.execute("insert into Faculty values(20201,'Vijay',25000)")
    conn.commit()
    print("Data Inserted")
except cx_Oracle.DatabaseError as e:
    if conn:
        conn.rollback()
        print("Connectivity problem",e)
finally:
    if csr:
        csr.close()
    if conn:
        conn.close()
```

Program: Write some code to update faculty salaries according to a range, such as increment by 600 whose salary is greater than 8000.

```
import cx_Oracle
try:
    conn=cx_Oracle.connect('scott/tiger@localhost')
    csr=conn.cursor()
    update_sal=float(input("Enter Update Salary:"))
    limit_sal=float(input("Enter Salary Limit:"))
    query="update Faculty set Fac_sal=Fac_sal+%f where Fac_sal<%f"
    csr.execute(query %(update_sal,limit_sal))
    print("Data Updated")
    conn.commit()
except cx_Oracle.DatabaseError as e:
    if conn:
        conn.rollback()
        print("onnectivity problem",e)
finally:
    if csr:
        csr.close()
    if conn:
        conn.close()
```

Program: Write some code to delete faculty salaries according to a range, such as delete faculties whose salary is greater than 8000.

```python
import cx_Oracle
try:
    conn=cx_Oracle.connect('scott/tiger@localhost')
    csr=conn.cursor()
    up_sal=float(input("Enter Salary:"))
    sql="delete from Faculty where Fac_sal>%f"
    csr.execute(sql %(up_sal))
    print("Data Deleted")
    conn.commit()
except cx_Oracle.DatabaseError as e:
    if conn:
        conn.rollback()
    print("Connectivity problem",e)
finally:
    if csr:
        csr.close()
    if conn:
        conn.close()
```

Program: Write some code to select all the faculty information using the fetchone() method.

```python
import cx_Oracle
try:
    conn=cx_Oracle.connect('scott/tiger@localhost')
    csr=conn.cursor()
    csr.execute("select * from Faculty")
    f=csr.fetchone()
    while f is not None:
        print(f)
        f=csr.fetchone()
except cx_Oracle.DatabaseError as e:
    if conn:
        conn.rollback()
        print("Connectivity Problem",e)
finally:
    if csr:
        csr.close()
    if conn:
        conn.close()
```

16.3.3 Working with the MySQL Database

In MySQL, logical databases are required, i.e. to work with one's own database. The default databases of MySQL are mysql, test, information_schema, and performance_schema.

The basic commands of MySQL are:

1. To know about the databases available

Syntax: show databases;

2. Creation of logical database

Syntax: create database dbname;

3. Dropping of database

Syntax: drop database dbname;

4. Using the specified logical database

Syntax: connect dbname;

5. Creation of a table

Syntax: create table faculty(f_no int(10) primary key, fname varchar(10));

6. Insertion of data

Syntax: insert into faculty values (20, 'Parth');

Program: Write a program using the MySQL database to create a table, and insert data into it and display it.

```
import mysql.connector
try:
    conn=mysql.connector.connect(host='localhost',database='Department',
                            user='root',password='root')
    csr=conn.cursor()
    csr.execute("create table Fac(Fac_id int(6) primary key,Fac_name varchar(8),
                Fac_sal double(8,2))")
    print("Table Done")
    query = "insert into Fac(Fac_id, Fac_name, Fac_sal) VALUES(%s, %s, %s)"
    info=[(20201,'Vijay',25000),(20202,'Swati',50000),(20203,'Parth',30000)]
    csr.executemany(query,info)
    conn.commit()
    print("Data Inserted")
    csr.execute("select * from Fac")
    d=csr.fetchall()
    for r in d:
        print("Id:",r[0])
        print("Name:",r[1])
        print("Salary:",r[2])
        print()
        print()
except mysql.connector.DatabaseError as e:
    if conn:
        conn.rollback()
        print("Connectivity problem",e)
finally:
    if csr:
        csr.close()

    if conn:
        conn.close()
```

Program: Write a program to copy the faculty table data from the MySQL database into the Oracle database.

```python
import mysql.connector
import cx_Oracle
try:
    conn=mysql.connector.connect(host='localhost',database='Department',
                                 user='root',password='root')
    csr=conn.cursor()
    csr.execute("select * from Fac")
    info=csr.fetchall()
    l=[]
    for r in info:
        tp=(r[0],r[1],r[2])
        list.append(tp)
except mysql.connector.DatabaseError as e:
    if conn:
        conn.rollback()
        print("Connectivity problem",e)
finally:
    if csr:
        csr.close()
    if conn:
        conn.close()
try:
    conn=cx_Oracle.connect('scott/tiger@localhost')
    csr=conn.cursor()
    query="insert into Fac values(:Fac_id,:Fac_name,:Fac_sal)"
    csr.executemany(query,l)

    conn.commit()
    print("Data Copied from MySQL to Oracle")
except cx_Oracle.DatabaseError as e:
    if conn:
        conn.rollback()
        print("Connectivity problem",e)
finally:
    if csr:
        csr.close()
    if conn:
        conn.close()
```

16.4 Conclusion

There are number of reasons for using Python in programming the database applications. Python program are much faster and efficient in comparison to other languages. It is well known for its portable nature, and it supports the SQL cursor and platform independence. In most of the programming languages, the developer has to take care of the opening and closing connection of the database to avoid further errors and exceptions, but in Python the PVM itself takes care of these connections.

Review Questions

1. How will you connect to MySQL from Python?
2. Mention the arguments needed to connect to MySQL from Python.
3. Explain exceptions related to databases.
4. Differentiate between fetchall() and fetchone().
5. What do we mean by commit and rollback?

Programming Assignments

PA 1: Create a table students with fields: sno: integer, sname: char, age: int with sno as the primary key.

- Insert values into students.
- Update students' age by 1.
- Delete students with age greater than 20.

Reference

Stones, Richard, and Neil Matthew. 2006. *Beginning databases with postgreSQL: From novice to professional*. Apress.

Appendix A: Mini Projects

1. Mail Sent

The main idea behind this mini project is to help you understand some basic concepts of Python such assmtplib module, for loop, f-strings, lists, and so on. Using these concepts we can develop a simple mail sending project that can send mail to multiple email addresses. When a user executes this project it asks the user about how many mail addresses they want to send this mail. Then it asks the user about the list of mail one by one. When the user has entered a complete list of mail, it asks the user about the login credentials through which the user wants to send this mail. If the user provides an invalid credential it terminates the application with an error message. If the login credential is valid then it asks the user about the subject and the body of the mail. When the user enters these it sends this mail to all the mail addresses that the user provided.

Code:

```
import smtplib

lst=[]
n=int(input("Enter number of email address you want to send mail"))
for i in range(n):
    email_address=input(f"enter {i+1} mail address : ")
    lst.append(email_address)

s = smtplib.SMTP('smtp.gmail.com', 587)
s.starttls()

your_email=input("Enter your email address")
your_password=input("Enter your email password")

s.login(your_email, your_password)

subject=input("Enter Subject of mail you want to send ")
text=input("Enter text of mail you want to send ")
```

```
message = 'Subject: {}\n\n{}'.format(subject, text)

for i in lst:
    s.sendmail(your_email, i, message)

s.quit()
print("Mail sent")
```

```
Enter number of email address you want to send mail3
enter 1 mail address : samarth.anand.cs.2018@miet.ac.in
enter 2 mail address : vijay.sharma@miet.ac.in
enter 3 mail address : samarthanand275@outlook.in
Enter your email addresskeeptotalrecord@gmail.com
Enter your email passwordYour_Password
Enter Subject of mail you want to send Subject Comes Here
Enter text of mail you want to send (Text of Mail Comes Here......)
Mail sent
```

2.   Generating a Calendar

Generate a calendar of any arbitrary year using a calendar module with pre-defined functions.

```
import calendar
p = int(input("Enter the year"))
a=1
print("\n")
Cal = calendar.TextCalendar(calendar.SUNDAY)
i=1
while i<=12:
    Cal.prmonth(p,i)
    i+=1
```

```
Enter the year2021


     January 2021
Su Mo Tu We Th Fr Sa
                1  2
 3  4  5  6  7  8  9
10 11 12 13 14 15 16
17 18 19 20 21 22 23
24 25 26 27 28 29 30
31
     February 2021
Su Mo Tu We Th Fr Sa
    1  2  3  4  5  6
 7  8  9 10 11 12 13
14 15 16 17 18 19 20
21 22 23 24 25 26 27
28
      March 2021
Su Mo Tu We Th Fr Sa
    1  2  3  4  5  6
 7  8  9 10 11 12 13
14 15 16 17 18 19 20
21 22 23 24 25 26 27
28 29 30 31
```

```
        April 2021              July 2021              October 2021
Su Mo Tu We Th Fr Sa    Su Mo Tu We Th Fr Sa    Su Mo Tu We Th Fr Sa
            1  2  3               1  2  3                      1  2
 4  5  6  7  8  9 10     4  5  6  7  8  9 10     3  4  5  6  7  8  9
11 12 13 14 15 16 17    11 12 13 14 15 16 17    10 11 12 13 14 15 16
18 19 20 21 22 23 24    18 19 20 21 22 23 24    17 18 19 20 21 22 23
25 26 27 28 29 30       25 26 27 28 29 30 31    24 25 26 27 28 29 30
        May 2021             August 2021        31
Su Mo Tu We Th Fr Sa    Su Mo Tu We Th Fr Sa          November 2021
                  1      1  2  3  4  5  6  7     Su Mo Tu We Th Fr Sa
 2  3  4  5  6  7  8     8  9 10 11 12 13 14        1  2  3  4  5  6
 9 10 11 12 13 14 15    15 16 17 18 19 20 21     7  8  9 10 11 12 13
16 17 18 19 20 21 22    22 23 24 25 26 27 28    14 15 16 17 18 19 20
23 24 25 26 27 28 29    29 30 31                21 22 23 24 25 26 27
30 31                                           28 29 30
        June 2021           September 2021             December 2021
Su Mo Tu We Th Fr Sa    Su Mo Tu We Th Fr Sa    Su Mo Tu We Th Fr Sa
       1  2  3  4  5              1  2  3  4              1  2  3  4
 6  7  8  9 10 11 12     5  6  7  8  9 10 11     5  6  7  8  9 10 11
13 14 15 16 17 18 19    12 13 14 15 16 17 18    12 13 14 15 16 17 18
20 21 22 23 24 25 26    19 20 21 22 23 24 25    19 20 21 22 23 24 25
27 28 29 30             26 27 28 29 30          26 27 28 29 30 31
```

---

3.   Dice Roll Simulator

Dice are simple cubes with the numbers 1 to 6 written on its faces. A dice simulator is a simple computer model that can roll dice for us and gives a different number after every roll.

This project involves writing a program that simulates rolling the dice. When the program executes, it will arbitrarily selects a number between 1 and6. For this, you have to set the minimum and maximum number that the dice can produce. Generally, a die contains a minimum of 1 and a maximum of 6. A function is needed that randomly grabs a number within the given range and prints it.

Code:

```
import random
while True:
    print(''' A. roll the dice              B. exit        ''')
    inp = int(input("what you want to do\n"))
    if inp==1:
        num = random.randint(1,6)
        print(num)
    else:
        break
```

```
 A. roll the dice          B. exit
what you want to do
1
2
 A. roll the dice          B. exit
what you want to do
1
1
 A. roll the dice          B. exit
what you want to do
1
5
 A. roll the dice          B. exit
what you want to do
2
```

4.   Arithmetic Calculator

Create a simple calculator that can perform basic arithmetic operations like addition, subtraction, multiplication, or division depending upon the user input.

Code:

```python
def addition ():
    print("Addition")
    n = float(input("Enter the number: "))
    t = 0
    ans = 0
    while n != 0:
        ans = ans + n
        t+=1
        n = float(input("Enter another number (0 to calculate): "))
    return [ans,t]
def subtraction ():
    print("Subtraction");
    n = float(input("Enter the number: "))
    t = 0
    sum = 0
    while n != 0:
        ans = ans - n
        t+=1
        n = float(input("Enter another number (0 to calculate): "))
    return [ans,t]
def multiplication ():
    print("Multiplication")
    n = float(input("Enter the number: "))
    t = 0
    ans = 1
    while n != 0:
        ans = ans * n
        t+=1
        n = float(input("Enter another number (0 to calculate): "))
    return [ans,t]
def average():
    an = []
    an = addition()
    t = an[1]
```

```
        a = an[0]
        ans = a / t
        return [ans,t]
while True:
    list = []
    print(" Enter 'a' for addition")
    print(" Enter 's' for substraction")
    print(" Enter 'm' for multiplication")
    print(" Enter 'v' for average")
    print(" Enter 'q' for quit")
    c = input(" ")
    if c != 'q':
        if c == 'a':
            list = addition()
            print("Ans = ", list[0], " total inputs ",list[1])
        elif c == 's':
            list = subtraction()
            print("Ans = ", list[0], " total inputs ",list[1])
        elif c == 'm':
            list = multiplication()
            print("Ans = ", list[0], " total inputs ",list[1])
        elif c == 'v':
            list = average()
            print("Ans = ", list[0], " total inputs ",list[1])
        else:
            print ("Sorry, invalid character")
    else:
        break
```

```
Enter 'a' for addition
Enter 's' for substraction
Enter 'm' for multiplication
Enter 'v' for average
Enter 'q' for quit
```

```
      m
Multiplication
Enter the number: 5
Enter another number (0 to calculate): 2
Enter another number (0 to calculate): 0
Ans =  10.0  total inputs  2
 Enter 'a' for addition
 Enter 's' for substraction
 Enter 'm' for multiplication
 Enter 'v' for average
 Enter 'q' for quit
      v
Addition
Enter the number: 3
Enter another number (0 to calculate): 4
Enter another number (0 to calculate): 0
Ans =  3.5  total inputs  2
 Enter 'a' for addition
 Enter 's' for substraction
 Enter 'm' for multiplication
 Enter 'v' for average
 Enter 'q' for quit
      a
Addition
Enter the number: 5
Enter another number (0 to calculate): 9
Enter another number (0 to calculate): 0
Ans =  14.0  total inputs  2
 Enter 'a' for addition
 Enter 's' for substraction
 Enter 'm' for multiplication
 Enter 'v' for average
 Enter 'q' for quit
      q
```

5. Mad Libs Generator

The Mad Libs Generator teaches you to employ user-inputted data such as Mad Libs, which refer to a series of inputs that a user enters. The input from the user could be anything from an adjective, a verb, a verb, or anything. After entering all the inputs, the application takes all the data and arranges it to deploy a story template.

Code:

```
loop = 1
while (loop < 10):

    noun = input("Choose a noun: ")
    p_noun = input("Choose a plural noun: ")
    noun2 = input("Choose a noun: ")
    place = input("Name a place: ")
    adjective = input("Choose an adjective (Describing word): ")
    noun3 = input("Choose a noun: ")

    print ("----------------------------------------------")
    print ("Be kind to your",noun,"He is very", p_noun)
    print ("For someone it may be somebody's", noun2,",")
    print ("Be kind to your",p_noun,"in",place)
    print ("Where the nature is always",adjective,".")
    print ()
    print ("You may think that it is the",noun3,",")
    print ("Well it is.")
    print ("----------------------------------------------")

    loop = loop + 1
```

```
Choose a noun: Parth
Choose a plural noun: good
Choose a noun: boy
Name a place: work
Choose an adjective (Describing word): best
Choose a noun: perfect
-----------------------------------------

Be kind to your Parth He is very good
For someone it may be somebody's boy ,
Be kind to your good in work
Where the nature is always best .

You may think that it is the perfect ,
Well it is.
-----------------------------------------
```

6. Speech to Text in Hindi

The main idea behind this mini project is to help you understand some basic concepts of Python such as the speech_recognition module, how to access the microphone, text conversion to Hindi words, and so on. Using these concepts, we can develop a simple Hindi speech recognition application. When the user executes this application it asks the user to say something, and when the user speaks it records the user's words and displays a "Done!" message on the screen. Then it converts the audio file to a Hindi text file through Google speech recognition and displays the text on the screen.

Code:

```
import speech_recognition as sr

r = sr.Recognizer()

with sr.Microphone() as source:
    print('Say Something:')
    r.adjust_for_ambient_noise(source, duration=0.2)
    audio = r.listen(source)
    print ('Done!')

text = r.recognize_google(audio, language = 'hi-IN')

print (text)

print (r.recognize_google(audio))
```

```
Say Something:
Done!
हेलो मेरा नाम समर्थ है मैं बीटेक कंप्यूटर साइंस का छात्र हूं
hello mera naam Samarth hai main BTech computer science
ka Chhatra hun
```

7.　Speech to Text and Text to Speech in English

The main idea behind this mini project is to help you understand some basic concepts of Python such as the speech_recognition and pyttsx3 module, how to access the microphone, function, while loop, exception handling, text conversion from audio to text, and so on.

Using these concepts, we can develop a simple speech-to-text and text-to-speech application. When we execute this application it displays a "Speak!" message on the screen and starts recording the user's audio. If it finds the audio file is empty it displays a "Can't hear" message on the screen and again starts recording the audio. When it finds text in the audio file, it starts converting audio to text through Google speech recognition and displays it on the screen. Then it sends the text to the speak function, which takes text as an input and converts it into audio and starts playing the audio. When the speak function has completely played the audio file then it returns to the while loop and again asks for speech. Since this program is in an infinite loop it continues to execute until it finds the word "exit" in the audio file. In the audio file, when it encounters the word "exit" the while loop gets terminated.

Code:

```python
import speech_recognition as sam
import pyttsx3

r = sam.Recognizer()

def speak(command):
    engine = pyttsx3.init()
    engine.say(command)
    engine.runAndWait()

while(1):
    try:

        with sam.Microphone() as source:
            print("Speak!")
            r.adjust_for_ambient_noise(source, duration=0.2)
            audio = r.listen(source)
            text = r.recognize_google(audio)
            text = text.lower()

            print("Did you say "+text)

            if("exit" in text):
                print("Thanks for using our Application")
                speak("Thanks for using our Application")
                break

            speak(text)

    except sam.RequestError as e:
        print("Could not request results; {0}".format(e))

    except sam.UnknownValueError:
        print("Can't hear")

Speak!
Did you say hello my name is samarth anand
Speak!
Did you say i am pursuing btech in computer science branch
from meerut institute of engineering and technology
Speak!
Can't hear
Speak!
Can't hear
Speak!
Did you say please exit the application
Thanks for using our Application
```

8.   Searching Using Python: Binary Search

Binary search is a technique that is used to search in a sorted array by repeatedly dividing the search interval in two halves. If the value of the search key is less than the item in the middle of the interval, decrease the interval to the lower half. Otherwise, decrease it to the upper half.

- Repeatedly check until the value is found or the interval is empty.
- We ignore half of the elements, after one comparison.
- Compare p with the middle element.
- If p matches with the middle element, we return the middle index.
- Else if p is greater than the middle element, then p can only lie in the right half of the sub array after the middle element. So we repeat it for the right half.
- Else (p is smaller) repeat for the left half.

Recursive solution:

```python
def binary_search(array, x):
    low = 0
    high = len(array) - 1
    mid = 0

    while low <= high:

        mid = (high + low) // 2

        if array[mid] < x:
            low = mid + 1

        elif array[mid] > x:
            high = mid - 1

        else:
            return mid

    return -1

array = [ 2, 3, 4, 10, 40 ]
p = 10

output = binary_search(arr, p)

if result != -1:
    print("Element is present at index", str(output))
else:
    print("Element is not present in array")
```

```
Element is present at index 3
```

Iterative solution:

```python
def binary_search(array, x):
    low = 0
    high = len(array) - 1
    mid = 0

    while low <= high:

        mid = (high + low) // 2

        if array[mid] < x:
            low = mid + 1

        elif array[mid] > x:
            high = mid - 1

        else:
            return mid

    return -1

array = [ 2, 3, 4, 10, 40 ]
p = 10

output = binary_search(arr, p)

if result != -1:
    print("Element is present at index", str(output))
else:
    print("Element is not present in array")
```

Element is present at index 3

Appendix B: Socket Programming in Python

A socket is the endpoint of a two-way communication link between two programs running on a network. A socket is bound to a port number so that the TCP layer can predict the application that the data is destined to be sent to. An endpoint is the combination of an IP-address and a port number. All the TCP connection can be individually predicted by its two endpoints. Thus, one can have multiple associations between the host and the server. Generally, a server executes on a specific system and has a socket that is bound to a specific port number. The server just waits, and listens to the socket for a client to make a connection request. The client knows the hostname of the system on which the server is executing and the port number on which the server is listening. To make an association request, the client tries to rendezvous with the server on the server's machine and the port as shown in Figure 1.

FIGURE 1
Connection request

If everything executes perfectly, the server accepts the connection request as in Figure 2.

FIGURE 2
Connection established

On the client side, if the connection request is granted, the socket is successfully established and the client can use the socket to connect with the server. The client and the server can now connect by writing or reading from their individual sockets. To create a socket, the socket module needs to be imported. The socket.socket() function is used to create a socket.

Syntax: s = socket.socket(socket_family, socket_type, protocol = 0)

Example:

```
import socket
s = socket.socket(socket.AF_INET, socket. SOCK_STREAM)
```

Here, a socket instance is made and two parameters are passed. The first parameter AF_ INET refers to the address family IPV4 and the second parameter SOCK_STREAM is meant for the connection-oriented TCP protocol. Table 1 represents different socket methods with their descriptions.

TABLE 1
Socket methods

Method	Description
s.bind()	It binds the address(hostname, portnumber) to socket.
s.listen()	It sets up and starts TCP listener.
s.accept()	It accepts TCP client connections, waiting until connection arrives.
s.connect()	It is a client socket method.
s.recv()	It receives TCP messages.
s.send()	It transmits TCP messages.
s.recvfrom()	It receives UDP messages.
s.sendto()	It transmits UDP messages.
s.close()	It closes the socket.
socket.gethostname()	It returns the host name.

Client: The socket.connect(hostname, port) opens a TCP connection to the hostname on the port. Once the socket is opened, you can read from it like an IO object.

Example:

```
import socket
s=socket.socket()
port=12345
s.connect(('127.0.0.1',port))
print(s.recv(1024))
s.close()
```

Connecting to a server: One can connect to a server only by knowing its IP address. If any error occurs during the creation of a socket then a socket error is thrown.

Example:

```
import socket
ip=socket.gethostbyname('www.google.com')
print(ip)
```

```
172.217.160.164
```

A server has a bind() method that binds it to the specific IP and port so that it can listen to incoming requests on that IP and port. Next, a server has a listen() method, which puts the server into the listen mode. This allows the server to listen to incoming connections. Lastly, a server has the accept() and close() method. The accept method initiates a connection with the client and the close method closes the connection with the client.

Example:

```python
import socket
s=socket.socket()
print('Socket Created')
port=12345
s.bind(('',port))
print("Socket binded to %s" %(port))
s.listen(5)
print('Socket is listening')
while True:
    c,addr = s.accept()
    print('Got Connection from',addr)
    c.send('Thank you for connecting')
    c.close()
```

```
Socket Created
Socket binded to 12345
Socket is listening
```

Appendix C: Parallel Processing in Python

Parallel processing is a way in which a task is executed concurrently in multiple processors on the same machine. It is designed to minimize the overall execution time. However, there is an overhead when communicating between the processes, which can eventually increase the overall time taken for small tasks instead of decreasing it. In Python, the multiprocessing component is preferred to execute independent parallel processes by using subprocesses. It allows you to strengthen different processors on a single machine; thus, the tasks can be executed in completely discrete memory. The maximum number of processes executed at a time is restrained by the number of processors that can be computed using the cpu_count() function.

```
import multiprocessing as mp
print("Number of processors: ", mp.cpu_count())
```

```
Number of processors:  8
```

Types of execution in parallel processing: synchronous and asynchronous execution

In synchronous execution, the processes are concluded in the same order in which they were started, which is attained by locking the main program until the corresponding processes are executed whereas in asynchronous execution it doesn't entail locking. As an output, the sequence of the output can become intermingled but is performed more quickly.

Example: Count how many numbers exist between a given range in each row.

Given a two-dimensional array, count how many numbers there are between a given range in each row.

```
import numpy as np
from time import time
np.random.RandomState(100)
Array = np.random.randint(0, 10, size=[200000, 5])
xyz = Array.tolist()
xyz[:5]
```

```
[[8, 9, 7, 1, 1],
 [8, 9, 3, 8, 6],
 [8, 7, 7, 8, 2],
 [7, 7, 7, 4, 4],
 [8, 5, 0, 1, 4]]
```

Without parallelization

```
def howmany_within_range(row, mini, maxi):
    add = 0
    for n in row:
        if mini <= n <= maxi:
            add = add + 1
    return add

results = []
for row in data:
    results.append(howmany_within_range(row, mini=4, maxi=8))

print(results[:10])
```

```
[2, 3, 4, 5, 3, 1, 3, 4, 1, 3]
```

Synchronous execution using parallelization

The basic method to parallelize any task is to take a particular function that should be run several times and make it execute in parallel on several processors. For this, pool function needs to be initialized with n number of processors and pass the function which needs to be parallelized.

multiprocessing. Pool() provides the apply(), map() and starmap() methods to make any function run in parallel.

apply(): It takes an args argument which accepts the parameters passed to the 'function-to-be-parallelized' as an argument.

```
import multiprocessing as mp
out = mp.Pool(mp.cpu_count())
results = [out.apply(howmany_within_range, args=(row, 4, 8)) for row in data]
out.close()
print(results[:10])
```

```
[3, 1, 4, 4, 4, 2, 1, 1, 3, 3]
```

map(): It takes only one iterable as an argument.

```
import multiprocessing as mp
def howmany_within_range_rowonly(row, mini=4, maxi=8):
    total = 0
    for n in row:
        if mini <= n <= maxi:
            total = total + 1
    return total
out = mp.Pool(mp.cpu_count())
results = out.map(howmany_within_range_rowonly, [row for row in data])
out.close()
print(results[:10])
```

```
[3, 1, 4, 4, 4, 2, 1, 1, 3, 3]
```

starmap(): It takes only one iterable as argument, but each element in that iterable should also be iterable.

```
import multiprocessing as mp
out = mp.Pool(mp.cpu_count())
output = out.starmap(howmany_within_range, [(row, 4, 8) for row in data])
out.close()
print(output[:10])
```

[3, 1, 4, 4, 4, 2, 1, 1, 3, 3]

Asynchronous execution using parallelization

The asynchronous execution analogue apply_async(), map_async() and starmap_async()
by starting the next process as soon as the earlier one finishes without considering the ini-
tial order. As an output, there is no assurance that the output will be of the same sequence
as the input.

apply_async() with callback function: It is similar to apply() besides the callback function,
which lets to know how the computed outputs should be saved.

```
import multiprocessing as mp
out = mp.Pool(mp.cpu_count())
output = []
def howmany_within_range2(i, row, mini, maxi):
    total = 0
    for n in row:
        if mini <= n <= maxi:
            total = total + 1
    return (i, total)
def collect_result(result):
    global output
    results.append(result)
for i, row in enumerate(data):
    pool.apply_async(howmany_within_range2, args=(i, row, 4, 8), callback=collect_result)
out.close()
out.join()
output.sort(key=lambda x: x[0])
output_final = [r for i, r in results]
print(output_final[:10])
```

[3, 1, 4, 4, 4, 2, 1, 1, 3, 3, 3]

apply_async() with callback function: It can be used without the callback function, a list of
pools will be produced as the output.

```
import multiprocessing as mp
out = mp.Pool(mp.cpu_count())
output = []
result_objects = [pool.apply_async(howmany_within_range2, args=(i, row, 4, 8))
                  for i, row in enumerate(data)]
output = [r.get()[1] for r in result_objects]
out.close()
out.join()
print(output[:10])
```

[3, 1, 4, 4, 4, 2, 1, 1, 3, 3]

Pool.starmap_async()

```
import multiprocessing as mp
out = mp.Pool(mp.cpu_count())
output = []
output = pool.starmap_async(howmany_within_range2, [(i, row, 4, 8)
                    for i, row in enumerate(data)]).get()
out.close()
print(output[:10])
```

[3, 1, 4, 4, 4, 2, 1, 1, 3, 3]

Index

Note: Page numbers in *italics* indicate a figure and page numbers in **bold** indicate a table on the corresponding page.